Metallic Glasses

Edited by Dragica Minić and Milica Vasić

Published in London, United Kingdom

IntechOpen

Supporting open minds since 2005

Metallic Glasses
http://dx.doi.org/10.5772/intechopen.77589
Edited by Dragica Minić and Milica Vasić

Contributors

Milica Vasić, Dušan Minić, Dragica Minić, Shank Kulkarni, Dharmendra Singh, Devinder Singh, R.S. Tiwari, Kiran Mor, Biswajit Swain, Swadhin Kumar Patel, Ajit Behera, Soumya Mohapatra

First published in London, United Kingdom, 2020 by IntechOpen
IntechOpen is the global imprint of INTECHOPEN LIMITED, registered in England and Wales, registration number: 11086078, 7th floor, 10 Lower Thames Street, London, EC3R 6AF, United Kingdom
Printed in Croatia

British Library Cataloguing-in-Publication Data
A catalogue record for this book is available from the British Library

Additional hard and PDF copies can be obtained from orders@intechopen.com

Metallic Glasses
Edited by Dragica Minić and Milica Vasić
p. cm.
Print ISBN 978-1-78985-487-9
Online ISBN 978-1-78985-488-6
eBook (PDF) ISBN 978-1-83880-074-1

We are IntechOpen,
the world's leading publisher of
Open Access books
Built by scientists, for scientists

4,600+

Open access books available

119,000+

International authors and editors

135M+

Downloads

Our authors are among the

151

Countries delivered to

Top 1%

most cited scientists

12.2%

Contributors from top 500 universities

Interested in publishing with us?
Contact book.department@intechopen.com

Numbers displayed above are based on latest data collected.
For more information visit www.intechopen.com

Meet the editors

Professor Dr Dragica Minić was born in a family of an Orthodox priest, in the wonderful town of Brus at the foot of the Kopaonik mountain area. There she received her elementary as well as high school education and Christian upbringing. She began her career as a professor of science at Brus Gymnasium and continued at the Belgrade University, Faculty for Physical Chemistry, having passed all titles from assistant to full professor, lecturing different courses in physical chemistry. During the rich pedagogical work, she has mentored more than hundred graduates, around thirty master and twenty PhD students. She has published more than 160 scientific papers and has taken part in more than 200 national and international scientific conferences. She has written several university books, chapters in scientific editions of international significance, and scientific monographs. She has continued to work actively in science even after retirement in 2016.

Dr. Milica Vasić is a research associate at Belgrade University. She was born in Belgrade, Serbia, where she received elementary education and a university diploma. She acquired additional skills in material science by attending training courses in Serbia and abroad, and she received several acknowledgements for achievements during her education. She completed her PhD degree in Physical Chemistry at Belgrade University, where she has been employed since 2011. Her research work is in the field of physical chemistry of materials and it is mainly focused on thermal stability, mechanism, and kinetics of thermally induced structural transformations in amorphous alloys. Beside the research activities, she regularly takes part in activities aimed to promote science.

Contents

Preface

The term "metallic glasses" is widely used to denote the amorphous alloys obtained by rapid quenching techniques. These materials, characterized by short range atom ordering, have found application in a variety of modern industries due to their favorable magnetic, electrical, mechanical, and anti-corrosion properties.

Kinetic and thermodynamic metastability is one of the main characteristics of metallic glasses. Inevitably, these materials tend to transform to more stable forms by crystallization, under conditions of high temperature or pressure, or during prolonged application at moderate temperatures. Their thermal stability is mainly determined by their chemical composition, while the functional properties are determined by both the chemical composition and microstructure of the material.

Thermally induced microstructural transformations could result in deterioration or improvement of the functional properties, as a consequence of the formation of hybrid amorphous/nanocrystalline structure with appropriate crystalline volume fraction and diameter of the formed crystals. Accordingly, functional properties of metallic glasses as well as their thermal stability, mechanism, and kinetics of microstructural transformations, represent topics of considerable interest for the practical application of metallic glasses.

Dragica M. Minić and Milica M. Vasić
University of Belgrade,
Faculty of Physical Chemistry,
Serbia

Chapter 1

Introductory Chapter: Metallic Glasses

Dragica M. Minić and Milica M. Vasić

1. Introduction

Fast-growing technological development imposes a need for new functional materials with improved physical and mechanical properties. Since their first synthesis in 1960 [1], amorphous alloys, also known as metallic glasses, have been a focus of numerous investigations due to their advanced mechanical, electrical, magnetic, and anti-corrosion properties, related to their isotropic structure and short-range atomic arrangement [2–6].

Generally speaking, metallic glasses are multi-component systems involving different metals (M_I-M_{II}) or metal and non-metal, i.e., metalloid (M-NM) components [7–9]. For the M_I-M_{II} systems, the metals belong to the groups of transition, rare-earth or alkaline metals, or are uranium, neptunium, or plutonium [2, 10, 11]. The M-NM systems can be represented by the general formula $M_{75-85}NM_{15-25}$ (at.%), where M is one or more metal elements, usually the transition or noble one, and NM is one or more metalloid or non-metal elements, most commonly B, Si, Ge, C, or P.

The metallic glasses are solid materials without structural translational periodicity, characteristic for a crystalline structure. From the atomic aspect, the structure of metallic glasses is analogous to the structure of liquids, characterized by macroscopic isotropy, nonexistence of the long-range atomic ordering, but existence of a short-range ordering at the atomic level. The short-range ordering of the atoms means that each atom is surrounded by the same atoms positioned at similar distances, where the lines drawn between the atom centers form similar angles, as a consequence of chemical bonds keeping the atoms together in solid state. Variation in inter-atomic distances and angles means the variation in the strength of chemical bonds, causing the softening of material in defined temperature interval instead of melting at defined temperature [12].

The ability of a liquid alloy to transform into the metallic glass is called the glass-forming ability (GFA). The GFA is determined by structural, thermodynamic, and kinetic parameters characterizing the system, i.e., chemical composition, geometrical arrangement of atoms, bonding and atomic size effects, cooling rate, and crystallization kinetics [5]. So far, many empirical criteria were proposed with the aim of predicting and explaining the GFA [5, 13–15]. The empirical criteria for easier glass formation can be expressed in five points as follows:

1. alloy is multi-component containing at least three elements, two of which are metals;

2. atomic radii difference among the three constituent elements should be at least 12%;

3. heats of mixing among the main three elements should be negative;

4. total content of non-metals (metalloids) amounts to around 20 at.%; and

5. heteronucleants (oxide crystal inclusions) must be removed.

Generally speaking, the metallic glasses are solid materials exhibiting all the important features of the solid state. However, the short-range ordered glassy structure is manifested by broad halo peaks in XRD patterns. Due to the macroscopic isotropy of amorphous materials, for the description of their atomic structure, radial distribution function can be used. It represents the average number density of atoms as a function of the distance from the chosen atom.

In order to explain the amorphous structure of metallic glasses, different models were proposed [16–20]. Bernal introduced the model of dense random packing of hard spheres (DRPHS) [16, 17], which includes the presence of only metal atoms in the structure. The Polk's modification of the Bernal's model positioned the metalloid atoms at the larger holes of the DRPHS structure, but gave satisfactory results only for B and C as non-metallic components [18]. On the other hand, according to Gaskell's model [19], the alloy structure is built from the ordered structural units composed of 200–400 atoms, identified as trigonal prisms, tetrahedra, or octahedra, forming random long-range structures. In spite of a relatively large number of the proposed models and their modifications, many details related to the structure of amorphous alloys still remain unclear.

The term "metallic glasses" denotes those amorphous alloys obtained by rapid quenching techniques. During fabrication of a glassy alloy, the crystallization, including the steps of nucleation and growth of the formed nuclei, must be avoided. This can be achieved in different ways, involving very fast cooling of an alloy melt, often at a rate of 10^6 K min^{-1}. The most frequently used amorphization procedures aimed at preparation of amorphous alloys include rapid quenching of a melt of appropriate chemical composition, most commonly on a cold rotating metal disc [21]. Cooling rate necessary for amorphization is determined by the chemical composition, i.e., by the nature of the components forming a melt [8, 14]. Other methods used to obtained amorphous alloys include vapor deposition [22], spray deposition [23], ion implantation [24], laser processing [25], chemical reduction [26], electrodeposition [27], mechanical alloying [28], etc.

Glassy state is structurally and thermodynamically metastable and prone to transformations under the conditions of elevated pressure or temperature, or even during prolonged usage at moderate temperature. They could occur through the processes of relaxation, partial or complete crystallization, and recrystallization, changing the microstructure of a material, providing a simple procedure for production of polycrystalline and composite materials with targeted properties. Crystallization process can be [6, 12]:

- polymorphous crystallization (amorphous phase transforms into a single crystalline phase without a change in composition);

- primary crystallization (composition of the first crystalline phase formed from the glass differs from that of the amorphous matrix, and then the crystals of the phase formed primarily serve as the sites of secondary and tertiary crystallization);

- eutectic crystallization (two different phases crystallize simultaneously, in a coupled fashion, and their overall composition does not differ from that of the glassy matrix).

The microstructural transformations show a significant impact on physical properties of the materials changing their functionality. Structural relaxation process preceding the crystallization, characteristic of metallic glasses, includes rearrangement of individual species on the atom level and decrease in free volume, changing the short-range order and influencing primarily their electrical and magnetic properties. Additionally, as a result of relaxation, density, elastic modulus, Curie temperature, and viscosity grow, while thermal resistivity, diffusivity, and fracture toughness decrease [12]. The relaxation process can be achieved by low-temperature annealing at temperatures below the crystallization temperature.

Partial crystallization of metallic glasses leads to the formation of nanostructured or composite materials, involving nanocrystals embedded in amorphous matrix, with specific physical properties. All these together make the metallic glasses extraordinary precursors for the production of materials with targeted functionality. Properties of metallic glasses and nanocrystalline alloys obtained from the amorphous precursors are determined by both, the alloy chemical composition and microstructure.

Almost all the glassy alloys with favorable magnetic properties contain a high percentage of transition metals or rare earth elements. In this sense, iron, cobalt, and nickel-based metallic glasses are soft magnetic materials. Their excellent combination of magnetic properties including low coercivity, relatively high saturation magnetization, zero magnetostriction as well as their relatively high electrical resistivity allows their application in transformer cores, magnetic sensors, magnetic shielding, amplifiers, information handling technologies [6, 29, 30], etc. On the other hand, addition of Nd and Pr provides their hard magnetic properties [31].

Metallic glasses are considered, from a mechanical point of view, very hard and strong materials, with high wear resistance [2, 6]. The high strength of these materials is a consequence of the fact that they do not contain defects characteristic for crystalline structure. Advantageous mechanical properties are exhibited by the multi-component alloys based on Ti, Zr, Al, Mg, Fe, Co, or Ni [5, 32–40]. However, these materials are characterized by limited plastic strain in tension, while the inhomogeneous deformation occurs through the formation of shear bands [6]. Fracture toughness of metallic glasses is somewhat lower than that of crystalline materials, but two orders of magnitude higher than in the case of oxide glasses [12]. Metallic glasses based on Al and Mg possess high specific strength, due to their low density and mass [39, 40]. As a result of their favorable mechanical properties, including high strength and large elastic elongation limit, metallic glasses are used in reinforcing composites, for sporting goods, microgears, aircraft parts, brazing foils [6, 12, 41, 42], etc.

Good corrosion resistance, observed for the metallic glasses containing Cr, Zr, Ni, Nb, Mo, or V, is a particularly important characteristic of these materials from the aspect of their applicability in modern technology [43–46]. Some metallic glasses are suitable for being used as biomedical materials (such as the TiZrCuPdSn alloys [47]), while some other glassy alloys show superconducting properties (such as the TiNb-based ones [48]).

From a technological point of view, nanocrystalline alloys obtained by partial crystallization of the glassy alloys represent a particularly interesting class of functional materials. The iron-based nanocrystalline alloys with the composition Fe-R-B (where R is rare earth element, B is boron) possess hard magnetic properties [49]. However, the soft magnetic materials in this class are nanocrystalline materials with the composition Fe-Si-B-Nb-Cu (FINEMET), Fe-M-B-Cu (M is Zr, Hb or F) (NANOPERM), Fe-Co-M-B-Cu (M is Zr, Hb or F) (HITPERM) [50], etc. To maintain favorable functional properties, in this case the soft magnetic ones, crystal size of the α-Fe or α-Fe(Si) in FINEMET or NANOPERM alloys must not exceed 15 nm [51]. To obtain nanocrystalline structure from the amorphous one, controlled fast nucleation and slow crystal growth are required. This can be achieved by an appropriate choice of the alloy composition and by thermal treatment as in the FINEMET-type alloys, where Cu is added to facilitate nucleation, while the Nb decreases the crystal growth rate [51–53].

In order to provide and maintain an amorphous or nanocrystalline structure of targeted functionality, thermal stability, thermodynamics, and kinetics of phase transformations thermally induced of amorphous and nanocrystalline materials should be known [8, 54–75]. This requires determination of the temperatures of all of the phase transformations as well as the kinetic triplets of these processes, consisting of Arrhenius parameters, activation energy, and pre-exponential factor, as well as kinetic model (conversion function). By determining the crystallization kinetic model, information about crystallization mechanism, including nucleation, crystal growth, and impingement effects can be obtained. In this way, the lifetime of specific microstructure, important for reliable applicability of materials, can be predicted.

Solid-state transformations are often complex processes, consisting of several concurrent or consecutive steps, manifested experimentally by compounded curve forms. In order to discuss all these steps and propose the most probable mechanisms, during the analysis, deconvolution of the compounded peaks (DSC, TG, or even XRD) by using different mathematical tools is required [76–84].

In view of the foregoing, metallic glasses have still been intriguing although studied for more than 50 years now, offering a wide range of practical applications either in the glassy or derivative form, and promising further technological improvement and development.

Author details

Dragica M. Minić* and Milica M. Vasić
Faculty of Physical Chemistry, University of Belgrade, Belgrade, Serbia

*Address all correspondence to: drminic@gmail.com

IntechOpen

References

[1] Klement W, Willens RH, Duwez P. Non-crystalline structure in solidified gold–silicon alloys. Nature. 1960;**187**: 869-870

[2] Suryanarayana C. Metallic glasses. Bulletin of Materials Science. 1984;**6**:579-594

[3] Flohrer S, Herzer G. Random and uniform anisotropy in soft magnetic nanocrystalline alloys. Journal of Magnetism and Magnetic Materials. 2010;**322**:1511-1514

[4] Zarebidaki A, Seifoddini A, Rabizadeh T. Corrosion resistance of $Fe_{77}Mo_5P_9C_{7.5}B_{1.5}$ in-situ metallic glass matrix composites. Journal of Alloys and Compounds. 2018;**736**:17-21

[5] Suryanarayana C, Inoue A. Iron-based bulk metallic glasses. International Materials Review. 2013;**58**:131-166

[6] Suryanarayana C, Inoue A. Metallic glasses. In: Ullmann's Encyclopedia of Industrial Chemistry. Weinheim: Wiley-VCH Verlag GmbH & Co. KGaA; 2012

[7] Hasegawa R, O'Handley RC, Tanner LE, Ray R, Kavesh S. Magnetization, magnetic anisotropy, and domain patterns of $Fe_{80}B_{20}$ glass. Applied Physics Letters. 1976;**29**:219-221

[8] Kaloshkin SD, Tomilin IA. The crystallization kinetics of amorphous alloys. Thermochimica Acta. 1996;**280/281**:303-317

[9] Mendelsohn L, Nesbitt E, Bretts G. Glassy metal fabric: A unique magnetic shield. IEEE Transactions on Magnetics. 1976;**12**:924-926

[10] Nassif E, Lamparter P, Sperl W, Steeb S. Structural investigation of the metallic glasses $Mg_{85.5}Cu_{14.5}$ and $Mg_{70}Zn_{30}$. A Journal of Physical Sciences. 1983;**38a**:142-148

[11] Korelis PT, Liebig A, Bjorck M, Hjorvarsson B, Lidbaum H, Leifer K, et al. Highly amorphous $Fe_{90}Zr_{10}$ thin films, and the influence of crystallites on the magnetism. Thin Solid Films. 2010;**519**:404-409

[12] Shiflet GJ, Leng Y, Hawk JW. Metallic glasses. In: Ullmann's Encyclopedia of Industrial Chemistry. Weinheim: Wiley-VCH Verlag GmbH & Co. KGaA; 2005

[13] Chen HS. Glassy metals. Reports on Progress in Physics. 1980;**43**:353-432

[14] Turnbull D. Under what conditions can a glass be formed. Contemporary Physics. 1969;**10**:473-488

[15] Egami T, Waseda Y. Atomic size effect on the formability of metallic glasses. Journal of Non-Crystalline Solids. 1984;**64**:113-134

[16] Finney JL. Short-range structure of amorphous alloys. Nature. 1979;**280**:847-847

[17] Bernal JD. The Bakerian Lecture, 1962, The structure of liquids. Proceedings of the Royal Society of London A. 1964;**280**:299-320

[18] Polk DE. The structure of glassy metallic alloys. Acta Metallurgica. 1972;**20**:485-491

[19] Gaskell PH. A new structural model for amorphous transition metal silicides, borides, phosphides and carbides. Journal of Non-Crystalline Solids. 1979;**32**:207-224

[20] Greer AL. Metallic glasses... on the threshold. Materials Today. 2009;**12**:14-22

[21] Budhani RC, Goel TC, Chopra KL, Chopra KL. Melt-spinning technique for preparation of metallic glasses. Bulletin of Materials Science. 1982;**4**:549-561

[22] Mader S, Nowick AS. Metastable Co-Au alloys: Example of an amorphous ferromagnet. Applied Physics Letters. 1965;**7**:57-59

[23] Moss M, Smith DL, Lefever RA. Metastable phases and superconductors produced by plasma-jet spraying. Applied Physics Letters. 1964;**5**:120-121

[24] Grant WA. Amorphous metals and ion implantation. Journal of Vacuum Science and Technology. 1978;**15**:1644-1649

[25] Breinan EM, Kear BH, Banas CM. Processing materials with lasers. Physics Today. 1976;**29**:44-50

[26] Wonterghem J, Mørup S, Koch CJW, Charles SW, Wells S. Formation of ultra-fine amorphous alloy particles by reduction in aqueous solution. Nature. 1986;**322**:622-623

[27] Eliaz N, Sridhar TM, Gileadi E. Synthesis and characterization of nickel tungsten alloys by electrodeposition. Electrochimica Acta. 2005;**50**:2893-2904

[28] Koch CC, Cavin OB, McKamey CG, Scarbrough JO. Preparation of "amorphous" $Ni_{60}Nb_{40}$ by mechanical alloy. Applied Physics Letters. 1983;**43**:1017-1019

[29] Hasegawa R. Soft magnetic properties of metallic glasses. Journal of Magnetism and Magnetic Materials. 1984;**41**:79-85

[30] Hernando A, Vázquez M, Barandiarán JM. Metallic glasses and sensing applications. Journal of Physics E. 1988;**21**:1129-1139

[31] Inoue A, Zhang T, Takeuchi A. Hard magnetic bulk amorphous alloys. IEEE Transactions on Magnetics. 1997;**33**:3814-3816

[32] Jang D, Gross CT, Greer JR. Effects of size on the strength and deformation mechanism in Zr-based metallic glasses. International Journal of Plasticity. 2011;**27**:858-867

[33] Inoue A, Zhang W, Zhang T. Thermal stability and mechanical strength of bulk glassy Ni–Nb–Ti–Zr Alloys. Materials Transactions. 2002;**43**:1952-1956

[34] Wang J, Li R, Xiao R, Xu T, Li Y, Liu Z, et al. Compressibility and hardness of Co-based bulk metallic glass: A combined experimental and density functional theory study. Applied Physics Letters. 2011;**99**:151911

[35] Ma C, Istihara S, Soejima H, Nishiyama N, Inoue A. Formation of new Ti-based metallic glassy alloys. Materials Transactions. 2004;**45**:1802-1806

[36] Blagojević VA, Vasić M, David B, Minić DM, Pizúrová N, Žák T, et al. Microstructure and functional properties of $Fe_{73.5}Cu_1Nb_3Si_{15.5}B_7$ amorphous alloy. Materials Chemistry and Physics. 2014;**145**:12-17

[37] Blagojević VA, Minić DM, Žák T, Minić DM. Influence of thermal treatment on structure and microhardness of $Fe_{75}Ni_2Si_8B_{13}C_2$ amorphous alloy. Intermetallics. 2011;**19**:1780-1785

[38] Minić DM, Blagojević VA, Minić DM, Gavrilović A, Rafailović L, Žák T. Influence of microstructure on microhardness of $Fe_{81}Si_4B_{13}C_2$ amorphous alloy after thermal treatment. Metallurgical and Materials Transactions A. 2011;**42A**:4106-4112

[39] Inoue A, Matsumoto N, Masumoto T. Al–Ni–Y–Co amorphous alloys with high mechanical strengths, wide supercooled liquid region and large glass-forming capacity. Materials Transactions, JIM. 1990;**31**:493-500

[40] Inoue A, Masumoto T. Mg-based amorphous alloys. Materials Science and Engineering A. 1993;**173**:1-8

[41] Dudina DV, Georgarakis K, Li Y, Aljerf M, LeMoulec A, Yavari AR, et al. A magnesium alloy matrix composite reinforced with metallic glass. Composites Science and Technology. 2009;**69**:2734-2736

[42] Ashby MF, Greer AL. Metallic glasses as structural materials. Scripta Materialia. 2006;**54**:321-326

[43] Tenhover MA, Lukco DB, Shreve GA, Henderson RS. Corrosion resistance of Cr-based amorphous metal alloys. Journal of Non-Crystalline Solids. 1990;**116**:233-246

[44] Qin FX, Zhang HF, Deng YF, Ding BZ, Hu ZQ. Corrosion resistance of Zr-based bulk amorphous alloys containing Pd. Journal of Alloys and Compounds. 2004;**375**:318-323

[45] Rife G, Chan PCC, Aust KT. Corrosion of iron-, nickel- and cobalt-base metallic glasses containing boron and silicon metalloids. Materials Science and Engineering. 1981;**48**:73-79

[46] Ma H, Wang W, Zhang J, Li G, Cao C, Zhang H. Crystallization and corrosion resistance of $(Fe_{0.78}Si_{0.09}B_{0.13})_{100-x}Ni_x$ (x = 0, 2 and 5) glassy alloys. Journal of Materials Science and Technology. 2011;**27**:1169-1177

[47] Zhu SL, Wang XM, Inoue A. Glass-forming ability and mechanical properties of Ti-based bulk glassy alloys with large diameters of up to 1 cm. Intermetallics. 2008;**16**:1031-1035

[48] Inoue A, Masumoto T, Suryanarayana C, Hoshi A. Superconductivity of ductile titanium-niobium-based amorphous alloys. Journal de Physique, Colloque. 1980;**41**:C8758-C8761

[49] Yang CJ, Ray R, O'Handley RC. Magnetic hardening in melt-spun Fe-R-B alloys. Materials Science and Engineering. 1988;**99**:137-141

[50] McHenry ME, Willard MA, Laughlin DE. Amorphous and nanocrystalline materials for applications as soft magnets. Progress in Materials Science. 1999;**44**:291-433

[51] Kulik T. Nanocrystallization of metallic glasses. Journal of Non-Crytalline Solids. 2001;**287**:145-161

[52] Hono K, Ping DH, Ohnuma M, Onodera H. Cu clustering and Si partitioning in the early crystallization stage of an $Fe_{73.5}Si_{13.5}B_9Nb_3Cu_1$ amorphous alloy. Acta Materialia. 1999;**47**:997-1006

[53] Zhang YR, Ramanujan RV. The effect of niobium alloying additions on the crystallization of a Fe–Si–B–Nb alloy. Journal of Alloys and Compounds. 2005;**403**:197-205

[54] Tkatch VI, Limanovskii AI, Kameneva VY. Studies of crystallization kinetics of $Fe_{40}Ni_{40}P_{14}B_6$ and $Fe_{80}B_{20}$ metallic glasses under non-isothermal conditions. Journal of Materials Science. 1997;**32**:5669-5677

[55] Xu D, Johnson WL. Crystallization kinetics and glass-forming ability of bulk metallic glasses $Pd_{40}Cu_{30}Ni_{10}P_{20}$ and $Zr_{41.2}Ti_{13.8}Cu_{12.5}Ni_{10}Be_{22.5}$ from classical theory. Physical Review B. 2006;**74**:024207

[56] Wang Y, Zhai H, Li Q, Liu J, Fan J, Li Y, et al. Effect of Co substitution for Fe on the non-isothermal crystallization kinetics of $Fe_{80}P_{13}C_7$ bulk metallic glasses. Thermochimica Acta. 2019;**675**:107-112

[57] Gavrilović A, Rafailović LD, Minić DM, Wosik J, Angerer P, Minić DM. Influence of thermal treatment on structure development and mechanical properties of amorphous $Fe_{73.5}Cu_1Nb_3Si_{15.5}B_7$ ribbon. Journal of Alloys and Compounds. 2011;**509S**:S119-S122

[58] Vasić MM, Surla R, Minić DM, Lj R, Mitrović N, Maričić A, et al.

Thermally induced microstructural transformations of $Fe_{72}Si_{15}B_8V_4Cu_1$ alloy. Metallurgical and Materials Transactions A. 2017;**48A**:4393-4402

[59] Gavrilović A, Minić DM, Rafailović LD, Angerer P, Wosik J, Maričić A, et al. Phase transformations of $Fe_{73.5}Cu_1Nb_3Si_{15.5}B_7$ amorphous alloy upon thermal treatment. Journal of Alloys and Compounds. 2010;**504**:462-467

[60] Blagojević VA, Vasić M, David B, Minić DM, Pizúrová N, Žák T, et al. Thermally induced crystallization of $Fe_{73.5}Cu_1Nb_3Si_{15.5}B_7$ amorphous alloy. Intermetallics. 2014;**45**:53-59

[61] Minić DM, Blagojević VA, Maričić AM, Žák T, Minić DM. Influence of structural transformations on functional properties of $Fe_{75}Ni_2Si_8B_{13}C_2$ amorphous alloy. Materials Chemistry and Physics. 2012;**134**:111-115

[62] Minić DM, Blagojević VA, David B, Pizúrová N, Žák T, Minić DM. Influence of thermal treatment on microstructure of $Fe_{75}Ni_2Si_8B_{13}C_2$ amorphous alloy. Intermetallics. 2012;**25**:75-79

[63] Minić DM, Gavrilović A, Angerer P, Minić DG, Maričić A. Structural transformations of $Fe_{75}Ni_2Si_8B_{13}C_2$ amorphous alloy induced by thermal treatment. Journal of Alloys and Compounds. 2009;**476**:705-709

[64] Minić DM, Blagojević VA, Minić DM, David B, Pizúrová N, Žák T. Nanocrystal growth in thermally treated $Fe_{75}Ni_2Si_8B_{13}C_2$ amorphous alloy. Metallurgical and Materials Transactions A. 2012;**43A**:3062-3069

[65] Minić DM, Minić DM, Žák T, Roupcová P, David B. Structural transformations of $Fe_{81}B_{13}Si_4C_2$ amorphous alloy induced by heating. Journal of Magnetism and Magnetic Materials. 2011;**323**:400-404

[66] Minić DM, Minić DG, Maričić A. Stability and crystallization of $Fe_{81}B_{13}Si_4C_2$ amorphous alloy. Journal of Non-Crystalline Solids. 2009;**355**:2503-2507

[67] Blagojević VA, Minić DM, Vasić M, Minić DM. Thermally induced structural transformations and their effect on functional properties of $Fe_{89.8}Ni_{1.5}Si_{5.2}B_3C_{0.5}$ amorphous alloy. Materials Chemistry and Physics. 2013;**142**:207-212

[68] Kuo YC, Zhang LS, Zhang WK. The crystallization kinetics of amorphous $Co_{78-x}Fe_xSi_8B_{14}$ ribbons. Journal of Applied Physics. 1981;**52**:1889-1891

[69] Yuan ZZ, Chen XD, Wang BX, Chen ZJ. Crystallization kinetics of melt-spun $Co_{43}Fe_{20}Ta_{5.5}B_{31.5}$ amorphous alloy. Journal of Alloys and Compounds. 2005;**399**:166-172

[70] Bayri N, Kolat VS, Izgi T, Atalay S, Gencer H, Sovak P. Crystallisation kinetics of $Co_{75-x}M_xSi_{15}B_{10}$ (M = Fe, Mn, Cr and x = 0, 5) amorphous alloys. Acta Physica Polonica A. 2016;**129**:84-87

[71] Jung HY, Stoica M, Yi S, Kim DH, Eckert J. Crystallization kinetics of $Fe_{76.5-x}C_{6.0}Si_{3.3}B_{5.5}P_{8.7}Cu_x$ (x = 0, 0.5, and 1 at. pct) bulk amorphous alloy. Metallurgical and Materials Transactions A. 2015;**46**:2415-2421

[72] Gharsallah HI, Sekri A, Azabou M, Escoda L, Suñol JJ, Khitouni M. Structural and thermal study of nanocrystalline Fe-Al-B alloy prepared by mechanical alloying. Metallurgical and Materials Transactions A. 2015;**46**:3696-3704

[73] Dong Q, Pan YJ, Tan J, Qin XM, Li CJ, Gao P, et al. A comparative study of glass-forming ability, crystallization kinetics and mechanical properties of $Zr_{55}Co_{25}Al_{20}$ and $Zr_{52}Co_{25}Al_{23}$ bulk metallic glasses. Journal of Alloys and Compounds. 2019;**785**:422-428

[74] Lozada-Flores O, Figueroa IA, Gonzalez G, Salas-Reyes AE. Influence of minor additions of Si on the crystallization kinetics of $Cu_{55}Hf_{45}$ metallic glasses. Thermochimica Acta. 2018;**662**:116-125

[75] Bizhanova G, Li F, Ma Y, Gong P, Wang X. Development and crystallization kinetics of novel near-equiatomic high-entropy bulk metallic glasses. Journal of Alloys and Compounds. 2019;**779**:474-486

[76] Vasić MM, Žák T, Pizúrová N, Roupcová P, Minić DM, Minić DM. Thermally induced microstructural transformations and anti-corrosion properties of $Co_{70}Fe_5Si_{10}B_{15}$ amorphous alloy. Journal of Non-Crystalline Solids. 2018;**500**:326-335

[77] Vasić MM, Blagojević VA, Begović NN, Žák T, Pavlović VB, Minić DM. Thermally induced crystallization of amorphous $Fe_{40}Ni_{40}P_{14}B_6$ alloy. Thermochimica Acta. 2015;**614**:129-136

[78] Vasić MM, Minić DM, Blagojević VA, Minić DM. Kinetics and mechanism of thermally induced crystallization of amorphous $Fe_{73.5}Cu_1Nb_3Si_{15.5}B_7$ alloy. Thermochimica Acta. 2014;**584**:1-7

[79] Blagojević VA, Vasić M, Minić DM, Minić DM. Kinetics and thermodynamics of thermally induced structural transformations of amorphous $Fe_{75}Ni_2Si_8B_{13}C_2$ alloy. Thermochimica Acta. 2012;**549**:35-41

[80] Vasić M, Minić DM, Blagojević VA, Minić DM. Mechanism and kinetics of crystallization of amorphous $Fe_{81}B_{13}Si_4C_2$ alloy. Thermochimica Acta. 2013;**572**:45-50

[81] Vasić M, Minić DM, Blagojević VA, Minić DM. Mechanism of thermal stabilization of $Fe_{89.8}Ni_{1.5}Si_{5.2}B_3C_{0.5}$ amorphous alloy. Thermochimica Acta. 2013;**562**:35-41

[82] Wang Y, Xu K, Li Q. Comparative study of non-isothermal crystallization kinetics between $Fe_{80}P_{13}C_7$ bulk metallic glass and melt-spun glassy ribbon. Journal of Alloys and Compounds. 2012;**540**:6-15

[83] Cole KM, Kirk DW, Singh CV, Thorpe SJ. Role of niobium and oxygen concentration on glass forming ability and crystallization behavior of Zr-Ni-Al-Cu-Nb bulk metallic glasses with low copper concentration. Journal of Non-Crystalline Solids. 2016;**445-446**:88-94

[84] Kotrlova M, Zeman P, Zuzjakova S, Zitek M. On crystallization and oxidation behavior of $Zr_{54}Cu_{46}$ and $Zr_{27}Hf_{27}Cu_{46}$ thin-film metallic glasses compared to a crystalline $Zr_{54}Cu_{46}$ thin-film alloy. Journal of Non-Crystalline Solids. 2018;**500**:475-481

Chapter 2

Metallic Glasses: A Revolution in Material Science

Swadhin Kumar Patel, Biswajit Kumar Swain,
Ajit Behera and Soumya Sanjeeb Mohapatra

Abstract

Metallic glasses represent one kind of advanced material, very popular in recent decades. These materials are very adaptable like plastics for their manufacturability in very complex shapes. TPF (Thermoplastic forming) based processes seem very good method to process them. These materials can compete with plastics but have metallic properties. They behave as magnetic materials with less hysteresis loss and less eddy current loss making them suitable for transformer and MEMS (Micro-Electromechanical System) applications. These materials exhibit good corrosion resistance, hardness and toughness. Based on the property and application, metallic glasses are good rivals to plastics, metals and ceramics. Chemical composition and kinetics of supercooling of these materials are the areas where young researchers can focus attention with a view to their improvement.

Keywords: metallic glass, crystalline and amorphous structure, supercooling, TPF-based processing

1. Introduction

In our day-to-day life, we use several types of products made from different materials. Based on the area of application, the desired properties of equipment for their production and processing can vary. According to this, material selection takes place in the way to supply the best possible outcomes compared to others. In recent decades, aluminum, steel and plastics have been the most commonly used materials. The aluminum is considered as the best choice for automobile and structural object for its low density and high specific strength, whereas, due to good strength and cost-effectiveness, steel is the most preferred material for structural applications like construction, railway industry etc. Likewise, plastics are used for various home and kitchen appliances and also for the interior design of buildings, vehicles etc. Plastics are very adaptable materials because of their easy processing, but characterized by low strength compared to metals. On the other hand, aluminum and steel lose the battle with plastics in area of their processing in order to produce very intricate shapes. Because of the high processing temperature, metal products may have more defects over plastics. In this sense, metallic glasses compete with both metals and plastics. These materials have good strength and toughness compared to plastics and can be formed in any desired intricate design compared to metals. Also they possess high corrosion and wear resistance. So, we can say that metallic glasses are the materials having the good properties of metals (like steel and aluminum) as well as good adaptability like plastics. In metallic glasses, during deformation, dislocation movements occurs, shearing a localization of atoms, but in the case of crystalline materials several defects occur weakening them [1].

As promising materials for different applications, metallic glasses are preferred over metals, ceramics, magnetic and some other types of existing materials due to their enhanced properties. Some of the important reasons for which we consider these glasses for specific applications are discussed in the followings.

1. As discussed earlier metallic glasses have no long-range of ordering like crystalline materials. It develops more homogeneity inside the material because defects like point defects, dislocations and stacking faults are absent.

2. These materials possess very high strength in the elastic region. It can be declared as a good yielding strength of the material which is higher than steel.

3. Because of the good homogeneity of atoms in metallic glasses, very good corrosion resistance is achieved along with good wear resistance.

4. Ordinary silica glasses are brittle in nature unlike the metallic glasses which are very tough materials.

5. These materials have good luster and mirror effects but they are opaque.

6. The metallic glasses are very hard materials and their fracture resistance is much better compared to ceramics.

7. Because of the metallic atoms, these glasses possess significant magnetic effects. It helps to easily magnetize or demagnetize these materials. Metallic glasses with soft magnetism have very small hysteresis loop. Due to the narrow hysteresis, in these glasses hysteresis loss is minimized.

8. Because of the amorphous structure of metallic glasses, their electrical resistivity is higher resulting in less eddy current loss during its application.

The good adaptation of these materials for different industrial applications is a consequence of combinations of those properties. All those applications are discussed later.

The discovery of metallic glass in 1960 motivated scientists to research and manufacture this kind of materials [2]. They were first manufactured in California Institute of Technology, USA. The researchers got the non-crystalline structure in Au-Si alloys. Rapid quenching of those alloys from their liquid state was conducted by the gun technology. They formed a very thin layer of metallic glass over a cold copper substrate. The reason to take copper as the substrate is because of its good thermal conductivity. After that, people are continuously discovering various metallic glasses with different compositions of elements [2]. After 2000 AD, people are making varieties of metallic glasses and its demand is increasing for industrial applications. In the 1990s, the development of different BMGs (Bulk Metallic Glasses) based on late transition metals (LTM) started. A. Inoue et al. successfully developed the Fe–Al–Ga–P–C–B BMGs in 1995 [3]. Today the availability and cost-effectiveness are the two major factors in selecting and production of such materials.

For the stabilization of a supercooled metallic liquid the three rules were proposed [4]:

1. multicomponent BMG systems must consist of at least three or more elements;

2. atomic size of the constituent elements should have a significant size difference (greater than 12%);

3. heat of mixing of the elements should be negative.

2. General description of metallic glasses

According to atomic arrangement, we can categorize the existing and man-made solid materials into two main groups: crystalline and amorphous. When there is a proper ordered arrangement of atoms then we say it is a crystalline material. If there is a random arrangement of atoms, then the material is called amorphous. The atomic arrangements of crystalline and amorphous materials are shown in **Figure 1**. To get such randomness, the sizes of the atoms are very important. Much difference in the atomic radius of the components leads to more randomness in the atomic arrangement. Glass forming is majorly concerned with the study of crystallization of materials in order to avoid crystallization. When metallic alloys are cooled at a very fast rate, possibilities of getting an ordered arrangement are poor [5].

The glass transition temperature (Generally denoted as "T_g") characterizes amorphous/glass nature of materials. This is more easily understood in the case of a polymer. If we cool a polymer from its liquid state, initially it undergoes cooling and it gets a rubbery state and then after crossing the T_g, it becomes brittle. This kind of phenomenon occurs in amorphous metals too. In case of metallic glasses, we can say that T_g is the temperature at which material gets soft from hard upon heating or get hard upon cooling. This definition for polymers and metals looks similar but it is restricted to amorphous and semicrystalline metals only. The best way to explain the process of getting an amorphous metal or metallic glass is by supercooling the metal from its liquid state. In **Figure 2** T_f is the freezing temperature. During cooling, the liquid goes beyond the freezing point and is known as supercooled metal which can have an amorphous structure [6]. In this way, we can get a metallic glass. In the absence of supercooling, the liquid has a tendency to crystallize [6].

During the formation of glass, the material should avoid the route of crystallization. Crystallization happens during the cooling of material below its liquidus temperature. The difference in Gibbs free energy between liquid and crystalline state is an important factor for the ability of a metal to crystallize or to become amorphous. Whenever there is a transformation between liquid to the solid-state of a material,

(a)   (b)   (c)

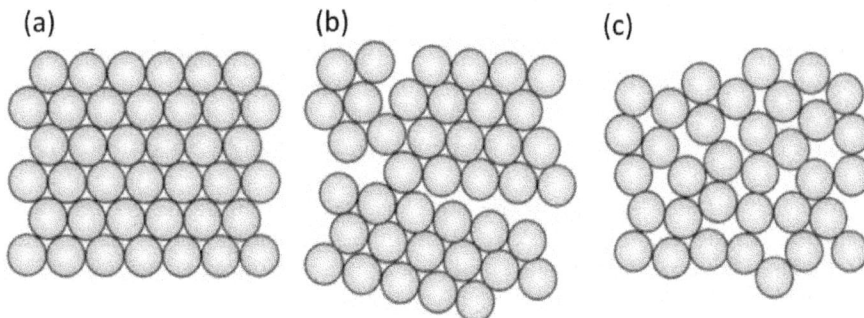

Figure 1.
The atomic arrangement of (a) single crystal, (b) polycrystal and (c) amorphous structures.

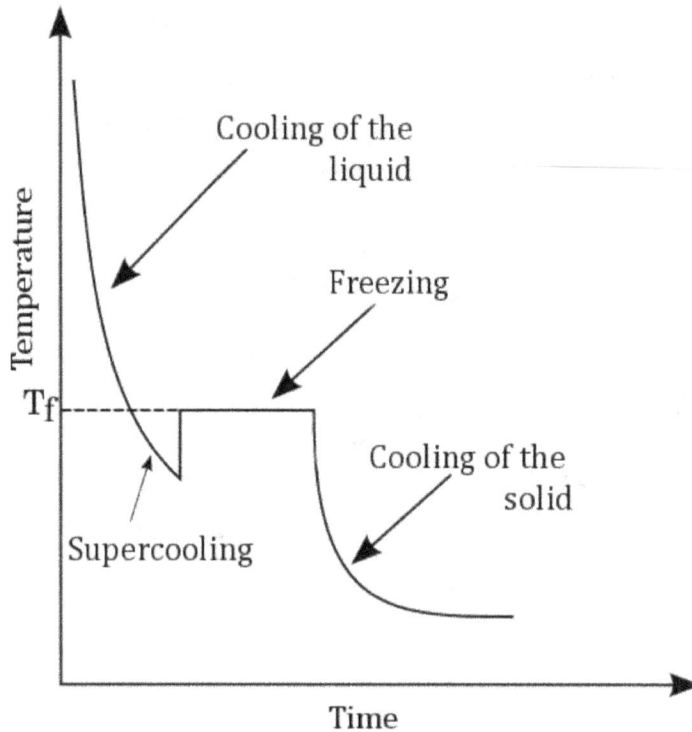

Figure 2.
Principle of supercooling.

the phase transformation at constant enthalpy gives a crystalline material. If enthalpy varies in that process, then the material escapes from the crystalline route and becomes amorphous. As we can see in **Figure 2**, a supercooling does not exist in the case of getting a crystalline material. On the other hand, in the case of supercooling, the enthalpy of transformation changes gradually. Therefore, during the manufacturing of the metallic glasses kinetics of the supercooling has a great impact on the quality of the glasses.

Considering the periodic table, the metallic glasses are mainly divided into two categories: metal–metal and metal–metalloid. In **Figure 3** elements which are metals and metalloids are shown in different colors for the better understanding of their selection for making metallic glasses.

As shown in the periodic table, metals are the elements starting from Lithium with atomic number 3. These are placed from group IA to VIA and shown by the yellow color in **Figure 3**. Some of the examples of metal–metal type metallic glasses are Ni–Nb, Mg–Zn, Hf–V, Cu–Zr, etc. Metals can be alkali, alkaline and rare earth metals etc. In metal–metal type, the atomic percentage of individual constituents can be up to 50%.

In the case of metal-metalloid type, one constituent is a metal and the other one is a metalloid. Metalloids like B, Si, Ge can be mixed with metals like Fe, Ni, Co etc. Metalloids in the periodic table are positioned in a step-like manner and colored with peach color. Metalloids have properties that are intermediate to both metals and non-metals. The composition percentage of metalloids in this category is lower than percentage of metals. After the discovery, the compositions of metalloids in the glasses were generally up to 20% but gradually people work on that problem and successfully decreased its percentage beyond 20% [7]. As described earlier atomic size, the heat of mixing is also considered for making metallic glasses and used for classification of these materials. The properties based on base metal and categories of metallic glasses are provided in **Tables 1** and **2**, respectively. This type of glasses is more often used in commercial applications.

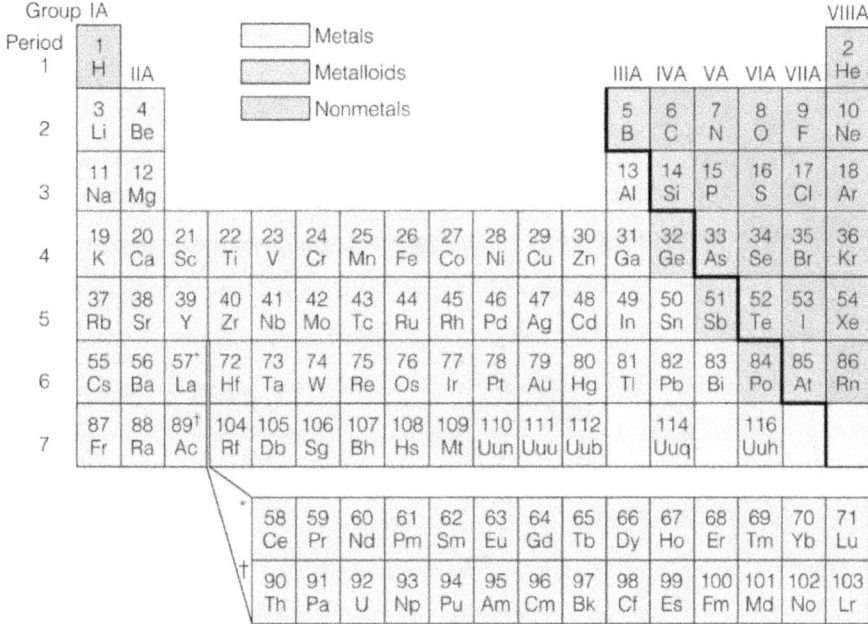

Figure 3.
Periodic table showing metals, metalloids and non-metals.

Base metal	Properties based on the base metal
Fe-based	Soft magnetism (glass, nanocrystal)
	Hard magnetism (nanocrystal)
	High corrosion resistance
	High endurance against cycled impact deformation
Co-based	Soft magnetism (glass, nanocrystal)
	Hard magnetism (nanocrystal)
	High corrosion resistance
	High endurance against cycled impact deformation
Ni-based	High strength, high ductility
	High corrosion resistance
	High hydrogen permeation
Cu-based	High strength, high ductility (glass, nanocrystal)
	High fracture toughness, high fatigue strength
	High corrosion resistance
Pd-based	High strength
	High fatigue strength, high fracture toughness
	High corrosion resistance

Table 1.
The basic properties of different LTM-based BMGs [8].

Base metal	Metal–Metalloids	Metal–Metal
Fe-based	Fe–(Al,Ga)–(P,C,B,Si) Fe–Ga–(P,C,B,Si) Fe–Ga–(Nb,Cr,Mo)–(P,C,B) (Los Alamos) Fe–(Cr,Mo)–(B,C) Fe–(Zr,Hf,Nb,Ta)–B Fe–(B,Si)–Nb	Fe–Nd–Al
Co-based	Co–Ga–(Cr,Mo)–(P,C,B) Co–(Zr,Hf,Nb,Ta)–B Co–Ln–B	Co–Nd–Al Co–Sm–Al
Ni-based	Ni–(NbCr,Mo)–(P,B) Ni–(Ta,Cr,Mo)–(P,B) Ni–Zr–Ti–Sn–Si (Yonsei University) Ni–Pd–P	Ni–Nb–Ti Ni–Nb–Zr Ni–Nb–Hf Ni–Nb–Zr–Ti Ni–Nb–Zr–Ti–M (M = Fe, Co, Cu) Ni–Nb–Hf–Ti Ni–Nb–Hf–Ti–M Ni–Nb–Sn (Cal Tech)
Cu-based	Cu–Pd–P Cu–Ni–Pd–P	Cu–Zr–Ti Cu–Hf–Ti Cu–Zr–Ti–Ni Cu–Hf–Ti–Ni Cu–Zr–Ti–Y Cu–Hf–Ti–Y Cu–Zr–Ti–Be Cu–Hf–Ti–Be Cu–Zr–Al Cu–Hf–Al Cu–Zr–Al–M Cu–Hf–Al–M (M = Ni, Co, Pd, Ag) Cu–Zr–Ga Cu–Hf–Ga Cu–Zr–Ga–M Cu–Hf–Ga–M Cu–Zr–Al–Y (Cal Tech)
Pd-based	Pt–Cu–P Pt–Cu–Co–P (Cal Tech) Pt–Pd–Cu–P	

Table 2.
Composition and properties of different types of BMGs (composition of base metal is greater than 50 at. %) [8].

3. Structure, properties and applications

Structure of material defines its property. BMGs do not exhibit a long-range order structure, as they solidify from liquid without reaching the crystalline ground state. However, short to medium-range structural order does develop to a considerable extent under the given kinetic constraints. This happens because the atoms strive to find comfortable configurations to lower their energy. The structure of the bulk metallic liquids was first observed by Bernal [9] and it was described as dense random packing. Structural features of metallic glasses are discussed by Michael et al. where the concept of efficient filling of space is supported [10]. The rationalization of the good glass forming compositions can be possible by the analysis of dense packing. An example of simple binary metallic glass is shown in **Figure 4** [10].

Figure 4.
Model of a simple binary metallic glass: Interpenetrating quasi-equivalent clusters sharing faces, edges, or vertices in the atomic packing configuration of a Zr-Pt metallic glass. The blue balls represent the solvent Zr atoms centered around Pt solute atoms [10].

These structural motifs arise from the strong tendency to form as many bonds as possible between unlike species because of the large negative heat of mixing which is usual in good glass formers. The size of the cluster and its type depend on the relative size of the solvent and the solute. The replacement of Pt solute in **Figure 4** by much smaller Be reduce the number of Zr neighbors which can be accommodated around the solute, and the solute concentration in the alloy would be correspondingly much higher. The medium-range order and dense packing in three-dimensional space can be possible by the overlapping of the cluster via various solvent-atom sharing schemes [11].

The adaptability of the metallic glass in the real world applications is spread in various fields, such as striking face plate in golf clubs, frame in tennis rackets, various shapes of optical mirrors, casing in cellular phones, casing in electro-magnetic instruments, connecting part of optical fibers, shot penning balls, electro-magnetic shielding plates, soft magnetic choke coils, soft magnetic high frequency power coils, high torque geared motor parts, high corrosion resistance coating plates, vessels for lead-free soldering, colliori type liquid flow meter, spring, in-printing plate, high frequency type antenna material, biomedical instruments such as endoscope parts etc. [12]. Metallic glasses are very strong compared to other conventional materials and that makes it a very good candidate for military applications like armor (Bulletproof vest) piercing bullets, anti-tank projectiles etc. Those metallic glasses, which are stronger than titanium, are also tried for aerospace application. Recently researchers from NASA and different research centers and organizations of China, Britain and Japan are doing several tests to get such ultimate material. This type of materials can give relatively double the performance compared to that of a titanium product in the space application. The major problem lies in the BMGs are the very quick aging of these materials. They become fragile after exposed to external physical stress conditions. The non-uniformity or not enough randomness inside the glass structure leads to the quick aging of these BMGs. So reliability cannot be achieved for the space applications. To remove such weakness of these materials, Wei-Hua Wang from China did experiments like the operation of a blacksmith. Cryogenic treatment of melted metal was done by using liquid nitrogen and maintained to room temperature after solidification. Again the material was melted and the process was repeated for several cycles. The purpose of choosing the repeated cryo-treatment is to reduce the instability of the material by increasing the randomness inside the material. This type of processing technique enhances the life of a BMG and increases its reliability.

Firstly China and secondly the United States are the major producers of BMGs. Currently, most applications are focused on electric based products like transformer core. Because of the good conductivity properties of BMGs, it dominates in that sector. Other applications like high-temperature applications, aerospace applications and military applications have a long way to go for becoming a better replacement for the recently used materials.

4. Processing of metallic glasses

The flow chart for the processing of metallic glasses in order to obtain different products is presented in **Figure 5**.

4.1 Direct casting

For the net shape fabrication of BMG, two types of casting processes (suction and die casting) have been adopted. Among the two methods the suction casting method develops a product with higher quality and lower porosity than die casting method. BMGs having low melting temperature are beneficial because this process reduces tool cost and wear, lowers energy consumption and shortens cycle time. The TTT diagram and the critical cooling rate (T_c) are highly affected by the maximum temperature prior to cooling rate [13]. The glass forming ability of the BMGs can be diminished if the maximum temperature prior to cooling does not exceed a threshold temperature (TOH). The overheating is rapid and it affects the viscosity in some type of BMGs [14]. This viscosity effect can be broadly explained by the phenomenon including the melting of an oxide phase, ordering phenomenon and a chemical decomposition process [15, 16]. During the casting processes (both die casting and suction casting) shrinkage has to be taken in to consideration. The shrinkage phenomenon is absent in the BMGs formers due to the absence of a first order phase transition during solidification. Cooling process in the casting plays a vital role. Low solidification shrinkage in BMGs develops a gap between the mold and the BMG during the cooling process [19, 20]. The heat transfer through the gap is different for the presence of atmosphere and vacuum, can become the rate

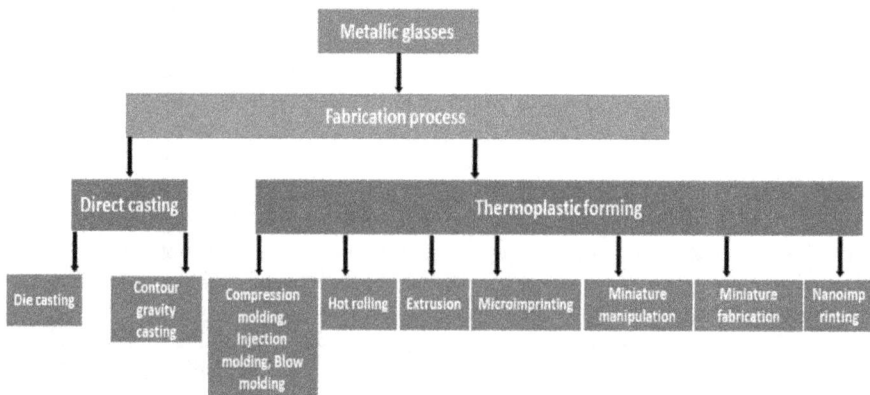

Figure 5.
Flow chart of the processing of metallic glasses.

limiting factor and leads to affect the cooling rate. In the process of direct casting of BMG formers, fast cooling and forming have to be done simultaneously due to the crystallization mechanism and the crystallization kinetics [7, 8]. Direct casting needs care during the mold filling and at the same time to avoid crystallization during solidification. This process is very difficult while making an intricate part which has attractive properties. **Figure 6** represents some application of BMGs fabricated by direct casting. The advantages and disadvantages of the direct casting BMG process are given in **Table 3**.

4.2 Thermoplastic forming

Thermoplastic forming (TPF) is the alternative process to direct casting for developing BMGs. This process has different nomenclature such as hot forming, hot pressing, superplastic forming, viscous flow working and viscous flow forming. The preferable condition for TPF is the drastic softening of the BMG former upon heating above T_g and its thermal stability. The measure of the ability of BMG formers to adopt an amorphous structure by heating above its glass transition temperature is known as thermal stability and it can be quantified by the width of the supercooled liquid region (SCLR). For $Zr_{44}Ti_{11}Cu_{10}Ni_{10}Be_{25}$ a very long processing time is available at low temperature by a high viscosity [17]. Furthermore, the viscosity is significantly reduced at a high temperature which leads to the reduction in processing time. Low viscosity and the long processing time are the

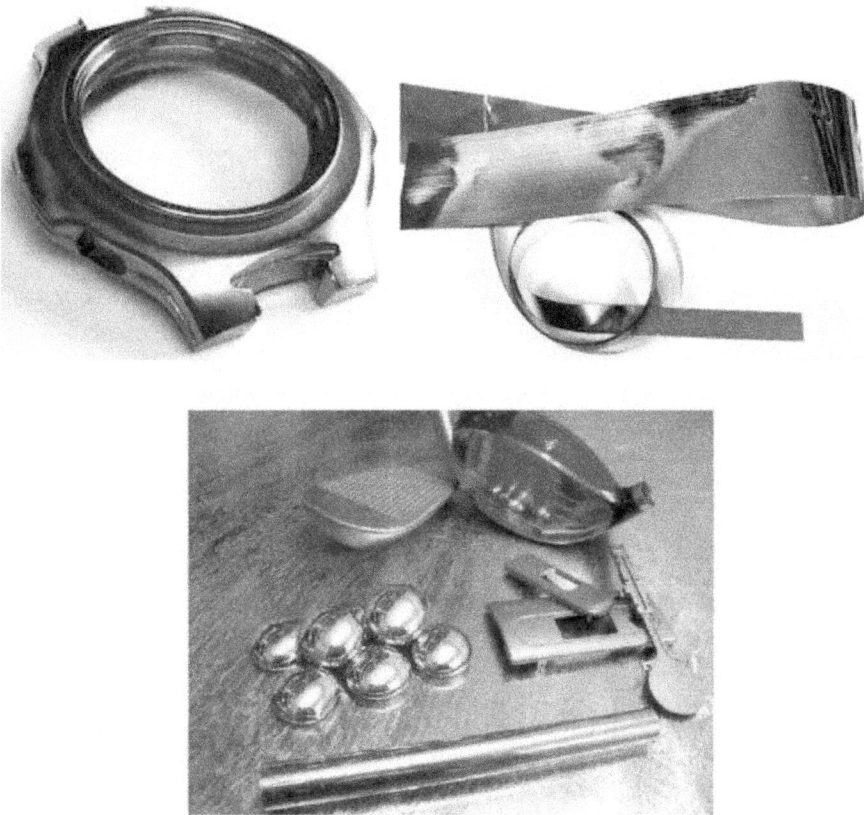

Figure 6.
BMG articles fabricated by direct casting method.

Advantages	Disadvantages
• Low melting temperature	• Cooling and forming are coupled
• Low shrinkage	• Processing environment can influence crystallization kinetics
• One step process	• High viscosity
• Homogeneous microstructure	• Internal stresses
• Mechanical properties are already matured in the as-cast state	• BMGs contaminate during processing

Table 3.
Advantages and disadvantages of the direct casting process.

optimum parameters for the highest formability of BMG former in its SCLR. According to the data reported in the open literature, the thermos physical properties that are detecting good formability of a BMG former are fragile liquid behavior, large poison ratio and low glass transition. Some TPF based BMG formers the glass transition temperature and the softening in the SCLR permits the use of processing temperature and pressure [18]. The capacity to plastically form metallic glasses in the zone of T_g was recognized by researchers. The enhanced formability of BMGs, which is developed by a wide range of processing methods based on the thermoplastic forming, attracts a wide range of researchers. In all the above-said processes, one thing is common that is the "feedstock". The shape and size of the feedstock are different for powder, over rods, disks, plates and films. These TPF based processing methods have been described in the below sections.

4.2.1 TPF based compression and injection molding

Compression molding was an adopted form of plastic processing. In this process, the feedstock material is placed in a mold and is given temperature into SCLR and pressure must exceed the flow stress of the BMG to achieve the required strain prior to crystallization setting. Fast cooling is not required for the forming. **Figure 7** represents the schematic diagram of the process equipped with examples. Under low applied pressure in the compression molding Pd-Ni-Cu-P alloy can be formed as a gear-shaped structure. It can be noticed that a dense compact part with mechanical properties close to those of bulk material and an outstanding surface finish can be obtained [11, 12]. **Figure 7** indicates compression molding of BM6 former, having different feedstock shapes such as pellets, plates and rods. The needed molding pressure not only depends on the formability of the BMG at the processing temperature but also on the shape of the final product. Injection molding is also a TPF based molding process. The only difference between the injection molding and compression molding is that in injection molding, feedstock material is rendered into the mold cavity which has the benefits for the development of the commercial fabrication processes with minimized cycle time.

4.2.2 Miniature fabrication

The development in technologies like micro-electromechanical systems (MEMS), electronics devices, and medical devices have created a rising demand for miniature products and parts. The miniature formation is done by different processes like German method LIGA (lithography, electroplating and molding), UV-LIGA etc. Due to the drawbacks of the LIGA i.e. cost, UV-LIGA is developed giving the similar products. Both processes can be used as the surface patterning

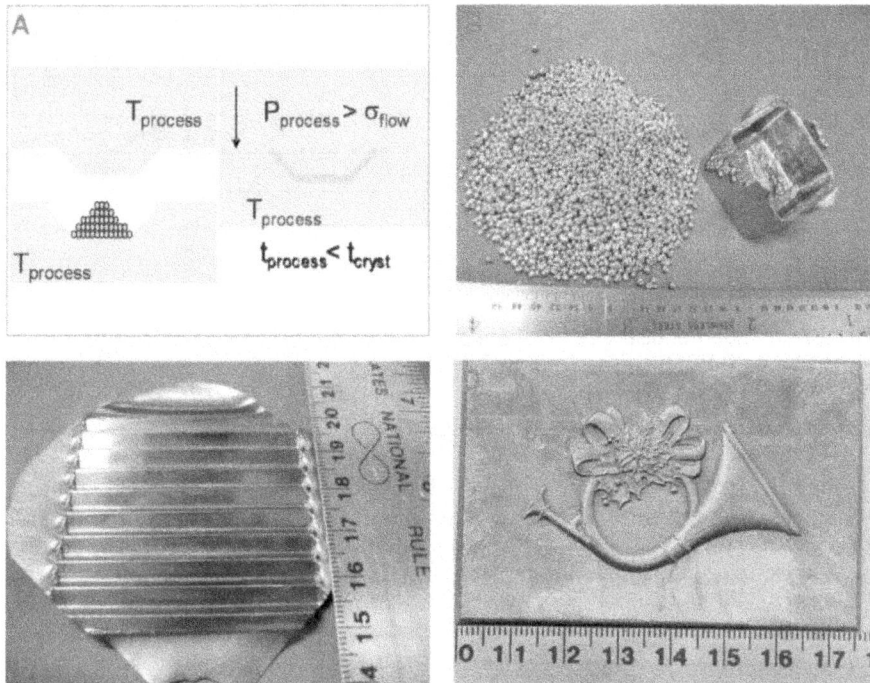

Figure 7.
TPF based compression molding with BMG. (A) Schematic diagram of the compression molding with BMGs. (B) Pellets used as feedstock material to compression mold $Pt_{57.5}Cu_{14.7}Ni_{5.3}P_{22.5}$. ($Zr_{44}Ti_{11}Cu_{10}Ni_{10}Be_{25}$) formed from a flat plate into a corrugated structure. (D) $Zr_{44}Ti_{11}Cu_{10}Ni_{10}Be_{25}$ formed from a flat plate to create an embossing mold.

techniques for creating high aspect ratio structures. Due to the homogeneous and isotropic structure of BMGs in atomic scale and their superior properties over conventional materials used for miniature applications and the capability to produce stress-free parts, these methods attracted a lot of attention.

4.2.3 Nano forming

BMGs basically formed by top-down nanofabrication. The combination of different properties like high strength at room temperature, the ability to imprint a nanometer-sized parallel print process, the non-linear softening of BMGs when reaching their glass transition and the ability to repeatedly write and erase facility on the BMG surface recommends a wide range of application. Nanoimprinting on BMG permits to directly write, such as with atomic force microscopy (AFM) tip, as in a scanning probe lithography process. The capability of BMGs for direct nanoimprinting can be applied with a combination of surface smoothening method and used as a rewritable high-density data storage. Several materials have been developed for the mold formation and imprint for nanoforming such as silicon, quartz, and alumina.

4.2.4 Rolling

Rolling of metallic glasses can be categorized into two processes; one is based on liquid processing and the other is on thermoplastic forming. The example of the former is the melt spinning. In the melt spinning, the liquid sample is quenched by

Advantages	Disadvantages
• Forming and fast cooling decoupled	• Two or more step process
• Highest dimensional accuracy	
• Insensitive to heterogeneous influence	• Novel and unique process
• Novel and unique process "green" process	
• Low capital investment	

Table 4.
Advantage and disadvantages of TPF based BMG processing.

injecting on a single, fast-spinning copper roll. Another process is cold rolling where the BMGs are rolled at the room temperature.

4.2.5 Extrusion

The thermoplastic formability of BMG formers can also be used for extrusion. The major advantage of extrusion is that it produces the highest aspect ratio shapes, 100+, of uniform cross-section [21]. During extrusion after the outlet of the effective die length, the swelling of the material is a common phenomenon (**Table 4**).

5. Conclusions

BMGs are now available in many different chemical compositions. Although various routes of their processing were adopted, still there is much scope for inventions targeting the development of new processing techniques and new types of glasses. TPF-based processing technique is accepted as a more suitable manufacturing process for obtaining glassy products. This process is competing with the processing of plastics. Very intricate shapes of different geometry, almost unachievable for metals, can be easily achieved with BMGs using TPF-based processing.

Author details

Swadhin Kumar Patel[1], Biswajit Kumar Swain[1*], Ajit Behera[1] and Soumya Sanjeeb Mohapatra[2]

1 Department of Metallurgical and Materials Engineering, NIT Rourkela, India

2 Department of Chemical Engineering, NIT Rourkela, India

*Address all correspondence to: enggbiswajit92@gmail.com

IntechOpen

References

[1] Schroers J. Processing of bulk metallic glass. Advanced Materials. 2010;**22**(14):1566-1597

[2] Klement WJ, Willens RH, Duwez P. Non-crystalline structure in solidified gold-silicon alloys. Nature. 1960; **187**(4740):869-870

[3] Inoue A, Shinohara Y, Gook JS. Thermal and magnetic properties of bulk Fe-based glassy alloys prepared by copper mold casting. Materials Transactions, JIM [Internet]. 2014;**36**(12):1427-1433. Available from: https://www.jstage.jst. go.jp/article/matertrans1989/36/12/ 36_12_1427/_article

[4] Inoue A. High strength bulk amorphous alloys with low critical cooling rates (Overview). Materials Transactions, JIM [Internet]. 2014; **36**(7):866-875. Available from: https:// www.jstage.jst.go.jp/article/matertra ns1989/36/7/36_7_866/_article

[5] Crystalline and Amorphous Solids: Explanation, Differences, Examples, etc [Internet]. 2019. Available from: https:// www.toppr.com/guides/chemistry/the- solid-state/crystalline-and-amorphous- solids/ [Cited: September 20, 2019]

[6] Busch R, Masuhr A, Johnson W. Pronounced asymmetry in the crystallization behavior during constant heating and cooling of a bulk metallic glass-forming liquid. Physical Review B: Condensed Matter and Materials Physics. 1999;**60**(17):11855-11858

[7] Chan KC, Sort J. Metallic glasses. Physical Metallurgy Fifth Edition. 2014; **1**(1):305-385

[8] Inoue A, Shen B, Takeuchi A. Developments and applications of bulk glassy alloys in late transition metal base system. Materials Transactions [Internet]. 2006;**47**(5):1275-1285.

Available from: http://joi.jlc.jst.go.jp/ JST.JSTAGE/matertrans/47.1275?from= CrossRef

[9] Bernal JD. Geometry of the structure of monatomic liquids. Nature [Internet]. 1960;**185**(4706):68-70. Available from: http://www.nature. com/articles/185068a0

[10] Miller M, Liaw P. Bulk metallic glasses: An Overview. 2007. 290 p

[11] Greer AL. Bulk Metallic Glasses: At the Cutting Edge of. MRS Bull. August 2007. 2019;**32**:611-619

[12] Zhang W, Inoue A, Wang XM. Developments and applications of bulk metallic glasses. Reviews on Advanced Materials Science. 2008;**18**:1-9

[13] Mukherjee S, Zhou Z, Schroers J, Johnson WL, Rhim WK. Overheating threshold and its effect on time–temperature-transformation diagrams of zirconium based bulk metallic glasses. Applied Physics Letters. 2004;**84**(24):5010-5012

[14] Lohwongwatana B, Schroers J, Johnson WL. Strain rate induced crystallization in bulk metallic glass- forming liquid. Physical Review Letters. 2006;**96**(7):1-4

[15] Ding S, Kong J, Schroers J. Wetting of bulk metallic glass forming liquids on metals and ceramics. Journal of Applied Physics. 2011;**110**(4):043508

[16] Kanik M, Bordeenithikasem P, Kumar G, Kinser E, Schroers J. High quality factor metallic glass cantilevers with tunable mechanical properties. Applied Physics Letters. 2014;**105**(13): 131911

[17] Nishiyama N, Inoue A. Glass- forming ability of Pd 42.5Cu 30Ni 7.5P 20 alloy with a low critical cooling rate

of 0.067 K/s. Applied Physics Letters. 2002;**80**(4):568-570

[18] Zong HT, Ma MZ, Wang LM, Liang SX, Liu RP. Effects of minor addition on glass forming ability: Thermal versus elastic criteria. Journal of Applied Physics. 2010;**107**(5):053515

[19] Gilman JJ. Mechanical behavior of metallic glasses. Journal of Applied Physics. 1975;**46**(4):1625-1633

[20] Patterson JP, Jones DRH. Moulding of a metallic glass. Materials Research Bulletin. 1978;**13**(6):583-585

[21] Sordelet DJ, Rozhkova E, Huang P, Wheelock PB, Besser MF, Kramer MJ, et al. Synthesis of $Cu_{47}Ti_{34}Zr_{11}Ni_8$ bulk metallic glass by warm extrusion of gas atomized powders. Journal of Materials Research [Internet]. 2002;**17**(1): 186-198. Available from: https://www.cambridge.org/core/article/synthesis-of-cu47ti34zr11ni8-bulk-metallic-glass-by-warm-extrusion-of-gas-atomized-powders/39E7068828966B930A0 3F1A7C457E31C

Thermal Stability and Phase Transformations of Multicomponent Iron-Based Amorphous Alloys

Milica M. Vasić, Dušan M. Minić and Dragica M. Minić

Abstract

Due to their excellent functional properties enabling their applicability in different fields of modern technology, amorphous alloys (metallic glasses) based on iron have been attracting attention of many scientists. In this chapter, the results of multidisciplinary research of five multicomponent iron-based amorphous alloys with different chemical composition, $Fe_{81}Si_4B_{13}C_2$, $Fe_{79.8}Ni_{1.5}Si_{5.2}B_{13}C_{0.5}$, $Fe_{75}Ni_2Si_8B_{13}C_2$, $Fe_{73.5}Cu_1Nb_3Si_{15.5}B_7$, and $Fe_{40}Ni_{40}P_{14}B_6$, are summarized in order to study the influence of chemical composition on their physicochemical properties and functionality. The research involved thermal stability, mechanism, thermodynamics, and kinetics of microstructural transformations induced by thermal treatment and their influence on functional properties. Determination of crystallization kinetic triplets of individual phases formed in the alloys is also included. The results obtained for different alloys are compared, correlated, and discussed in terms of the alloy composition and microstructure.

Keywords: amorphous alloys, iron, microstructure, crystallization, kinetics, functional properties

1. Introduction

Amorphous alloys (metallic glasses), composed of metallic and metalloid elements, characterized by a short-range atomic ordering, have been attracting a lot of scientific attention because of their extraordinary isotropic physical and mechanical properties [1–3]. Within this class of materials, the iron-based alloys stand out by a unique combination of magnetic, electrical, mechanical, and anticorrosion properties, which makes them suitable for many applications, as multifunctional materials [3–5]. Their applications as soft magnetic materials are mainly based on their low coercivity, high permeability, high saturation induction, low eddy current losses, low magnetic reversal losses, and high Curie temperature [6, 7]. Due to their high strength and hardness, large elastic elongation limit, and good corrosion resistance, amorphous alloys are convenient for different structural applications [3, 8]. Their functional properties as well as their thermal stability can be tuned by an appropriate choice of alloying elements. It is considered that the glass-forming ability of the alloy is improved if empirical component rules [9, 10] are fulfilled:

alloy should include more than three elements, metallic and nonmetallic, in the composition where the differences in atomic size of the three constituent elements are higher than 12%, negative heats of mixing among the main three constituents, a total amount of nonmetallic components of around 20 atomic %, and the absence of oxide inclusions. The alloys composed of more elements exhibit better glass-forming ability, which is known as "confusion principle" [11].

Thermodynamic metastability and kinetic metastability are among the key characteristics of amorphous alloys in general. Consequently, there is a high tendency for their transformations to more stable forms to occur under the conditions of elevated temperature and pressure or even during prolonged usage at moderate temperatures. These transformations include structural relaxation, glass transition, crystallization, and recrystallization processes, which affect the functional properties of the alloys, involving either their deterioration or improvement [12, 13]. When nanocrystals are formed in an amorphous matrix making a composite, the properties of the material are determined by crystal dimensions and volume fraction of the present nanocrystals. In the case of iron-based materials, the best hard magnetic properties can be obtained for full or almost full crystallization of the starting amorphous material, while the optimal soft magnetic properties can be achieved in the case of partial crystallization [14]. Accordingly, in order to tailor materials with targeted functional properties, information about thermal stability as well as the knowledge of mechanism and kinetics of thermally induced structural changes and their influence on functional properties of these materials are very important.

The goal of this chapter is to correlate and explain the results of our multidisciplinary studies [15–30] of five multicomponent iron-based amorphous alloys of different compositions, $Fe_{81}Si_4B_{13}C_2$, $Fe_{79.8}Ni_{1.5}Si_{5.2}B_{13}C_{0.5}$, $Fe_{75}Ni_2Si_8B_{13}C_2$, $Fe_{73.5}Cu_1Nb_3Si_{15.5}B_7$, and $Fe_{40}Ni_{40}P_{14}B_6$, in terms of mechanism, thermodynamics, and kinetics of thermally induced microstructural transformations.

2. Experimental

Iron-based amorphous alloys studied herein were prepared in the form of the 30–35 μm thin ribbons by melt-quenching technique [15–30]. The nominal composition of the as-prepared alloy samples can be represented as follows in atomic %: $Fe_{81}Si_4B_{13}C_2$, $Fe_{79.8}Ni_{1.5}Si_{5.2}B_{13}C_{0.5}$, $Fe_{75}Ni_2Si_8B_{13}C_2$, $Fe_{73.5}Cu_1Nb_3Si_{15.5}B_7$, and $Fe_{40}Ni_{40}P_{14}B_6$.

X-ray diffraction (XRD) measurements were performed in Bragg-Brentano geometry, using a Co K_α radiation source, at room temperature. Preparation of thermally treated samples included isothermal annealing of the alloy samples sealed in a quartz ampoule at selected temperatures, for 60 min in the case of the $Fe_{79.8}Ni_{1.5}Si_{5.2}B_{13}C_{0.5}$ alloy and for 30 min for all the other alloys. Qualitative and quantitative analyses of the collected XRD data of the as-prepared and thermally treated samples were conducted using ICSD [31], PDF-2 [32], and COD [33] databases and Maud [34] software.

Transmission electron microscopy (TEM) images were recorded with a Philips CM12 microscope (tungsten cathode, 120 kV electron beam). For TEM measurements, samples were prepared using the focused-ion beam (FIB) method (Ga ions). JEOL JSM 6460 was used to collect scanning electron microscopy (SEM) images.

Thermal analyses of the studied alloys were carried out by means of differential scanning calorimetry (DSC) in a protective nitrogen or helium atmosphere, at constant heating rates. Complex crystallization peaks were deconvoluted [19, 21–24] using either Gaussian-Lorentzian cross-product function or Fraser-Suzuki function, taking into consideration the criteria related to the nature of the process as well as the mathematical criteria.

Thermomagnetic measurements were conducted in an evacuated furnace using an EG&G vibrating sample magnetometer, under magnetic field of 4 kA m^{-1}, at constant heating rate. Electrical resistivity measurements were performed by the four-point method, in an inert atmosphere. Vickers microhardness was determined using MHT-10 (Anton Paar, Austria) microhardness testing device, with 0.4 N loads and 10 s loading time.

3. Results and discussion

Considering the metastability of amorphous alloys, preservation of microstructure and knowledge of thermal stability in wide temperature range are crucial for their practical applications. In this sense, our investigations start by structural characterization of several as-prepared Fe-based amorphous alloys of different chemical compositions, followed by thermal analysis.

3.1 Structural characterization of the as-prepared alloys

In order to obtain detailed information on microstructure of the as-prepared alloys and the nature of individual crystallization steps, the XRD and Mössbauer spectroscopy methods were applied [15, 20, 25, 28, 30]. The XRD results revealed that the microstructure of the as-prepared alloys is characterized by short-range atomic ordering showing characteristic broad diffraction halo maxima. According to the positions of broad diffraction halo maxima (2θ = 52 and 96°, **Figure 1**), the starting atomic configuration corresponds to the bcc-Fe structure, for all the studied alloys. Short-range ordering domain sizes for all the alloys were estimated to be approximately 1.6 nm according to the Scherrer equation [35].

Nevertheless, the as-prepared structures of the $Fe_{73.5}Cu_1Nb_3Si_{15.5}B_7$ and $Fe_{81}B_{13}Si_4C_2$ alloys are not completely amorphous, containing certain amounts of crystalline phases. Based on the sharp maximum in the XRD diagram (**Figure 1**) and the results of Mössbauer spectroscopy [28], 5% of the structure of the

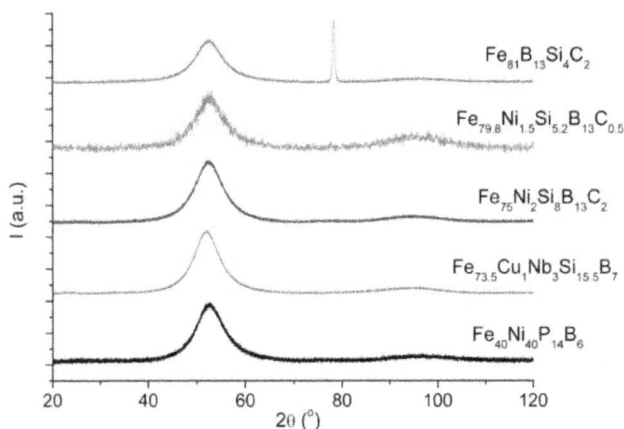

Figure 1.
XRD patterns of the as-prepared alloys.

as-prepared $Fe_{81}B_{13}Si_4C_2$ alloy is in crystalline form. This can be caused by high Fe content in this alloy and the fact that it does not contain any metal element other than Fe, so the requirements for easier amorphization [11] are not fully met. On the other hand, according to Mössbauer spectroscopy [20], 3.5% of the structure of the as-prepared $Fe_{73.5}Cu_1Nb_3Si_{15.5}B_7$ alloy correspond to crystalline clusters and disappear on heating, during the process of structural relaxation. This amount of crystalline phase is very close to the lowest amount which could be detected by XRD and consequently was not noticed in the XRD patterns (**Figure 1**). The appearance of crystallinity in this case was contributed by the presence of Cu atoms, which, when present in small amounts, form clusters serving as precursors for nucleation of the α-Fe(Si) crystalline phase.

3.2 Thermal stability of the alloys

According to the results of thermal analysis, all of the studied alloys possess good thermal stability at temperature under 380°C (**Figure 2a**). The glass transition preceding crystallization can be clearly observed only for the $Fe_{79.8}Ni_{1.5}Si_{5.2}B_{13}C_{0.5}$ and $Fe_{81}B_{13}Si_4C_2$ alloys (**Figure 2b**), suggesting their higher glass-forming ability than those of the other alloys studied. By applying DSC method, correlation between thermal stability of the alloys and their chemical composition was observed. The lowest thermal stability was demonstrated by the alloy with lower content of iron, $Fe_{40}Ni_{40}P_{14}B_6$, containing P instead of Si, with the beginning of crystallization at round 380°C. The alloys with higher content of iron show higher thermal stability, where the alloys $Fe_{73.5}Cu_1Nb_3Si_{15.5}B_7$, $Fe_{75}Ni_2Si_8B_{13}C_2$, and $Fe_{81}B_{13}Si_4C_2$ begin to crystallize at around 500°C and $Fe_{79.8}Ni_{1.5}Si_{5.2}B_{13}C_{0.5}$ even at around 520°C. The temperatures corresponding to the start of crystallization observed for the examined alloys (**Table 1**) are in agreement with the literature data for the similar systems [36]. A somewhat higher thermal stability of the $Fe_{79.8}Ni_{1.5}Si_{5.2}B_{13}C_{0.5}$ alloy was also suggested by a wide temperature range corresponding to supercooled liquid region (**Figure 2b**). This behavior results from the optimal chemical composition including two metal elements with the total content of around 80% (at.) and three nonmetallic amorphizers with the total content of around 20%.

The peak shape of exothermal stabilization maxima, sharp or rounded in some degree, and the presence of more than one maximum (**Figure 2**) indicate the occurrence of several parallel or consecutive steps of thermal stabilization, for all the studied alloys [19, 21–24]. The alloys containing higher amount of iron ($Fe_{79.8}Ni_{1.5}Si_{5.2}B_{13}C_{0.5}$ and $Fe_{81}B_{13}Si_4C_2$) exhibit one well-defined sharp crystallization peak, while the alloys with slightly lower amount of iron ($Fe_{73.5}Cu_1Nb_3Si_{15.5}B_7$ and $Fe_{75}Ni_2Si_8B_{13}C_2$) show two distinct completely separated compounded peaks (**Figure 2a**), which correspond to different crystallization and recrystallization steps. The alloy containing equal amounts of Fe and Ni (40% at.) exhibits two (one compounded and one sharp) partially overlapped DSC peaks, pointing out several crystallization steps.

The enthalpies of different crystallization steps for all the alloys are determined from the area corresponding to DSC peaks at various heating rates. Various heating rates yield different enthalpy values, showing that thermal history of a sample has a significant impact on the final state of the system. The starting state of the system is the same, but the final state is influenced by duration of thermal treatment as well as by the temperature, influencing the value of determined enthalpies. The observed average absolute values of the enthalpies at heating rates 5–20°C min^{-1} (**Table 1**) for crystallization are 80–110 J/g, but for recrystallization are around 20 J/g.

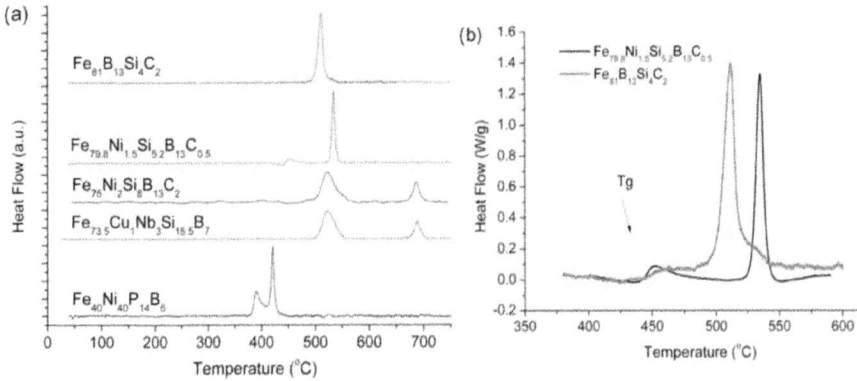

Figure 2.
DSC curves of the studied alloys at 5°C/min (a) and corresponding curves of the $Fe_{79.8}Ni_{1.5}Si_{5.2}B_{13}C_{0.5}$ and $Fe_{81}B_{13}Si_4C_2$ alloys in the temperature region 380–600°C indicating glass transition, T_g (b).

		T_0 (°C)	$|\Delta H|$ (J g^{-1})	T_{c1} (°C)	T_{c2} (°C)
$Fe_{81}B_{13}Si_4C_2$	α-Fe(Si), Fe₃B, Fe₂B	500	87	420	730
$Fe_{79.8}Ni_{1.5}Si_{5.2}B_{13}C_{0.5}$	α-Fe(Si), Fe₂B	520	110 ± 10	—	—
$Fe_{75}Ni_2Si_8B_{13}C_2$	Peak 1 α-Fe(Si), Fe₃B, Fe₂B	500	80 ± 20	430	740
	Peak 2	670	20 ± 6		
$Fe_{73.5}Cu_1Nb_3Si_{15.5}B_7$	Peak 1 α-Fe(Si), Fe₂B	500	90 ± 20	340	600
	Peak 2 Fe₁₆Nb₆Si₇, Fe₂Si	670	20 ± 10		
$Fe_{40}Ni_{40}P_{14}B_6$	α-(Fe,Ni), γ-(Fe,Ni), (Fe,Ni)₃(P,B)	380	—	360	480

Table 1.
Temperatures of the crystallization onset (T_0), average absolute values of the transformation enthalpies ($|\Delta H|$), and Curie temperatures (T_c), for individual amorphous alloys.

3.3 Thermally induced structural transformations

For more information on thermally induced microstructural transformation of the alloys and the nature of individual crystallization steps, the XRD, Mössbauer spectroscopy, and SEM and TEM methods were applied on the alloy samples isothermally treated at different temperatures, chosen according to the DSC thermograms.

With thermal treatment of the alloys, new narrow peaks appear in the XRD diffractograms as a result of crystallization. The changes of their relative intensities and areas point out the changes in microstructural parameters of the formed phases provoked by thermal treatment at different temperatures. The appearance and disappearance of some sharp peaks in the XRD patterns with a rise in temperature of thermal treatment indicate the processes of recrystallization and formation of one phase at the expense of another [15, 18, 20, 26, 30]. The analysis of the obtained XRD patterns yielded the information about microstructure of the studied samples and the phase composition diagrams (**Figure 3, Table 2**).

Due to the presence of bcc-Fe-like atomic configuration in the ordered domains of the as-prepared alloys, the α-Fe(Si) phase is the first crystalline phase formed in amorphous matrix during thermal treatment of the alloys [15, 18, 20, 26, 30].

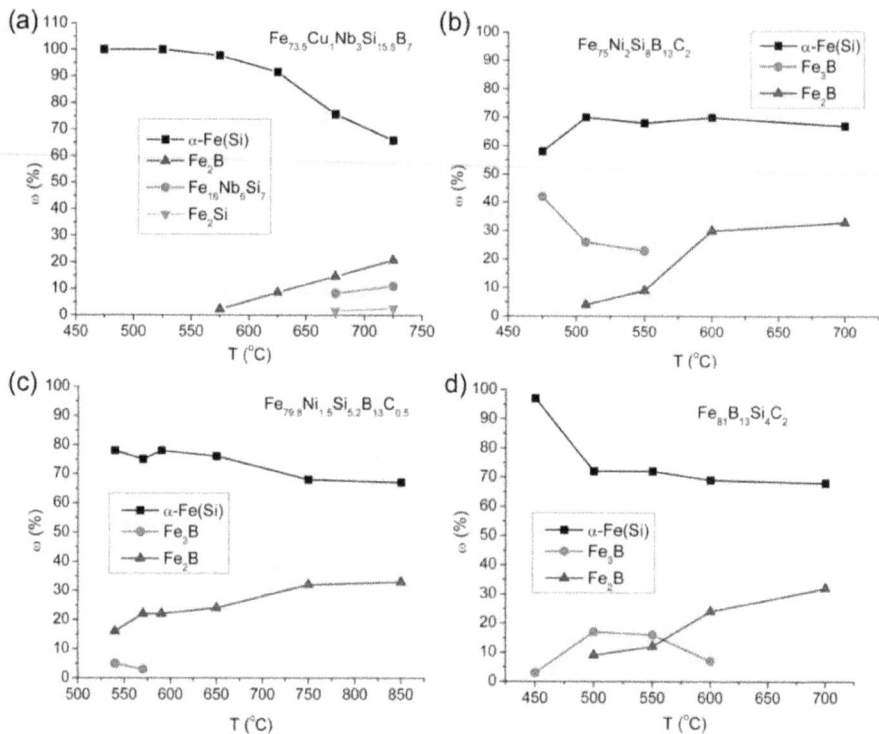

Figure 3.
Phase composition diagrams of the alloys containing 73–81 atomic % of iron ($Fe_{73.5}Cu_1Nb_3Si_{15.5}B_7$ (a), $Fe_{75}Ni_2Si_8B_{13}C_2$ (b), $Fe_{79.8}Ni_{1.5}Si_{5.2}B_{13}C_{0.5}$ (c), $Fe_{81}B_{13}Si_4C_2$ (d)), showing the fractions of individual phases relative to the total amount of the crystalline material in the alloy ((a) is reprinted from ref. [16] with permission of Institute of Physics of Polish Academy of Sciences).

Annealing temperature (°C)	Phases
340	α-(Fe,Ni); γ-(Fe,Ni)
370	α-(Fe,Ni); γ-(Fe,Ni); (Fe,Ni)$_3$(P,B)
400	α-(Fe,Ni); γ-(Fe,Ni); (Fe,Ni)$_3$(P,B)
420	α-(Fe,Ni); γ-(Fe,Ni); (Fe,Ni)$_3$(P,B)
500	γ-(Fe,Ni); (Fe,Ni)$_3$(P,B)
600	γ-(Fe,Ni); (Fe,Ni)$_3$(P,B)

Table 2.
Crystalline phases present in the $Fe_{40}Ni_{40}P_{14}B_6$ alloy after thermal treatment at different temperatures.

For these alloys, the beginning of crystallization of the α-Fe(Si) phase from the amorphous structure is observed at approximately the same temperatures (around 450–500°C), with the exception of the $Fe_{40}Ni_{40}P_{14}B_6$ alloy containing the lowest amount of iron (380°C). This phase remains the dominant crystalline phase over the whole temperature range examined. Formation of the α-Fe(Si) phase is also contributed by a tendency toward creation of stronger bonds between Fe and Si than between Fe and B, and repulsion between Si and B, as indicated by ab initio molecular dynamic simulations [37].

Crystallization of the α-Fe(Si) phase brings about favorable conditions for crystallization of boride phases, since in amorphous matrix, in the vicinity of

α-Fe(Si) grains, the ratio of boron to iron concentration is increased. This is contributed by several factors. Formation of α-Fe(Si) crystalline grains reduces Fe content in the amorphous matrix, while the boron is repulsed out of the α-Fe(Si) crystalline grains because of its low solubility in α-Fe and the presence of Si in this crystalline phase. Thus, amorphous/crystal interphase boundaries, being boron enriched, serve as nucleation sites for crystallization of boron phases. In the alloys containing 13 atomic % of boron ($Fe_{81}B_{13}Si_4C_2$, $Fe_{79.8}Ni_{1.5}Si_{5.2}B_{13}C_{0.5}$, and $Fe_{75}Ni_2Si_8B_{13}C_2$), two boride crystalline phases appear during heating: metastable Fe_3B and stable Fe_2B. In the case of $Fe_{73.5}Cu_1Nb_3Si_{15.5}B_7$, which contains 7 atomic % of boron, the metastable Fe_3B phase is observed only using Mössbauer spectroscopy and in lower amount than in the alloys containing higher quantity of boron [20]. Upon further heating, the metastable Fe_3B phase is transformed into the stable Fe_2B phase. The highest content of the Fe_3B phase is observed in the $Fe_{75}Ni_2Si_8B_{13}C_2$ alloy (**Figure 3**), which could be a consequence of the presence of a suitable amount of Ni in the alloy, since it is considered that the Ni element present in an appropriate amount can retard the degradation of metastable boride phases [38]. In the case of the $Fe_{79.8}Ni_{1.5}Si_{5.2}B_{13}C_{0.5}$ alloy, the Fe_3B phase can be observed only in very low amounts (few wt. % of crystalline phases) (**Figure 3c**), which is partially caused by longer heating time during sample preparation (60 min instead of 30 min). For all the alloys with Fe as the dominant component, crystalline phases α-Fe(Si) and Fe_2B are observed as final crystallization products [15, 16, 25, 28]. For the alloys containing 13 atomic % of boron, at the highest temperatures of thermal treatment, weight percentages of the α-Fe(Si) and Fe_2B crystalline phases are 70 and 30%, respectively, while, in the case of the $Fe_{73.5}Cu_1Nb_3Si_{15.5}B_7$ alloy, weight percentage of the Fe_2B phase at the highest temperatures of thermal treatment is lower (around 20% wt.), due to the lower boron content in the alloy. In addition, in this alloy, crystalline phases $Fe_{16}Nb_6Si_7$ and Fe_2Si are formed after heating at high temperatures [16], because of higher Si content than in the other alloys examined and the presence of Nb. Similarly to the alloys with Fe as the dominant component, crystallization of the $Fe_{40}Ni_{40}P_{14}B_6$ alloy starts with the formation of the bcc-structured phase, α-(Fe,Ni), but in this case it starts at lower annealing temperatures, 340–380°C. However, the crystallization mechanisms of the $Fe_{40}Ni_{40}P_{14}B_6$ alloy are somewhat different from those of the alloys containing 73–81% Fe and include the formation of crystalline phases α-(Fe,Ni), γ-(Fe,Ni), and (Fe,Ni)$_3$(P,B) and transformation of the α-(Fe,Ni) phase into γ-(Fe,Ni) and (Fe,Ni)$_3$(P,B) at high temperatures [18]. Actually, at higher temperatures, the crystalline phase with body centered cubic structure (α-(Fe,Ni)) is destabilized by high Ni content.

Application of TEM method confirms the results of XRD analysis and Mössbauer spectroscopy in terms of crystalline phases formed during heating [18, 20], showing that, after heating at the highest temperatures, the alloy structure is composed of grains, several 10s to several 100s of nanometers in size and irregular in shape, which are formed by coalescence of neighboring grains and influenced by impingement (**Figure 4**) [19–21]. Crystallization changes the morphology of the alloy sample and the distribution of individual elements on the surface of a sample [19, 20], which, after formation of crystalline phases, becomes nonuniform. As a result of crystallization, the alloy structure is more porous, because of imperfect packing of the crystals (**Figure 4b**) [17, 20, 26]. Surface morphology depends significantly on the heating rate and the temperature up to which the sample was heated, in other words on thermal history of a sample [19].

The thermal treatment causes continuous growth of the average crystallite size of α-Fe(Si) and Fe_2B phases in the alloys containing Fe as the dominant metal component except for the $Fe_{73.5}Cu_1Nb_3Si_{15.5}B_7$ alloy, according to the XRD analysis (**Figure 5**). However, it can be observed that the average crystallite size of α-Fe(Si) phase in the $Fe_{73,5}Cu_1Nb_3Si_{15,5}B_7$ alloy remains the same, around 15 nm, over a wide temperature range. This is expected as a consequence of the presence of Nb atoms in the amorphous matrix, which, due to their large atomic radius, hinder the diffusion of Fe and Si to the crystal, obstructing its further growth [14]. When crystallization of the phase containing Nb starts, further crystal growth of α-Fe(Si) phase occurs.

For all the alloys examined, the average crystallite size of the α-Fe(Si) phase at the highest temperatures amounts to 80–100 nm, except for the $Fe_{79,8}Ni_{1,5}Si_5$ $_{,2}B_{13}C_{0,5}$ alloy, where it is around 35 nm. This exception can originate from the crystallization kinetics of individual steps of formation of α-Fe(Si) phase in this alloy, where a higher ratio of the nucleation rate to the crystal growth rate than in the other alloys examined occurs. On the other hand, when it comes to another phase observed in all the alloys containing Fe as the dominant component, Fe_2B, its crystallite size reaches approximately 50 nm after heating at the highest temperatures, except for the $Fe_{73,5}Cu_1Nb_3Si_{15,5}B_7$ alloy, where the size of around 30 nm is contributed by lower boron content in the alloy. In the case of the alloy containing 40 atomic % of iron, in accordance with the chemical composition and unique phase compositions, during thermal treatment, the average crystallite size of the formed phases changes slightly in temperature ranges in which nucleation is the dominant process or exhibits more pronounced changes in temperature intervals where the crystal growth dominates [18].

3.4 Influence of thermal treatment on functional properties

Functional properties of the as-prepared and thermally treated amorphous alloys are significantly influenced by their microstructure beside the chemical composition. In the case of very low thermal effects, thermally induced microstructural transformations are more noticeable in the changes of functional properties than by thermal analysis. Bearing this in mind as well as potential practical application of the studied alloys, microhardness, thermomagnetic resistivity, and electrical resistivity analyses were performed.

3.4.1 Microhardness

In the as-prepared form, the examined alloys exhibit relatively high microhardness values, over 900 HV [20, 26, 30], as shown for the $Fe_{73,5}Cu_1Nb_3Si_{15,5}B_7$, $Fe_{75}Ni_2Si_8B_{13}C_2$, and $Fe_{81}B_{13}Si_4C_2$ alloys (**Figure 6a**). Thermally induced formation of nanocrystalline structure results in an increase in the microhardness value, which reaches maximum at around 500–600°C and then declines (**Figure 6a**). The

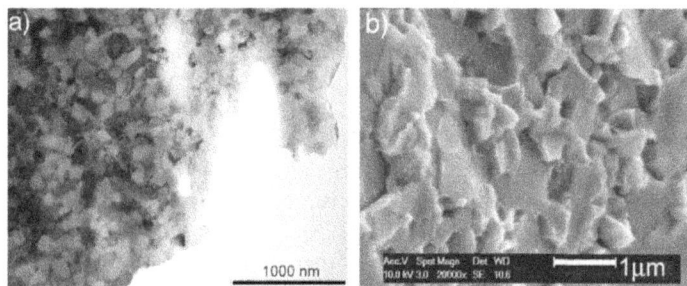

Figure 4.
TEM image of the $Fe_{73,5}Cu_1Nb_3Si_{15,5}B_7$ alloy sample annealed at 725°C (a) and SEM image of the cross section of the $Fe_{73,5}Cu_1Nb_3Si_{15,5}B_7$ alloy sample annealed at 850°C for 24 h (b) as the examples showing the microstructure of the crystallized alloy.

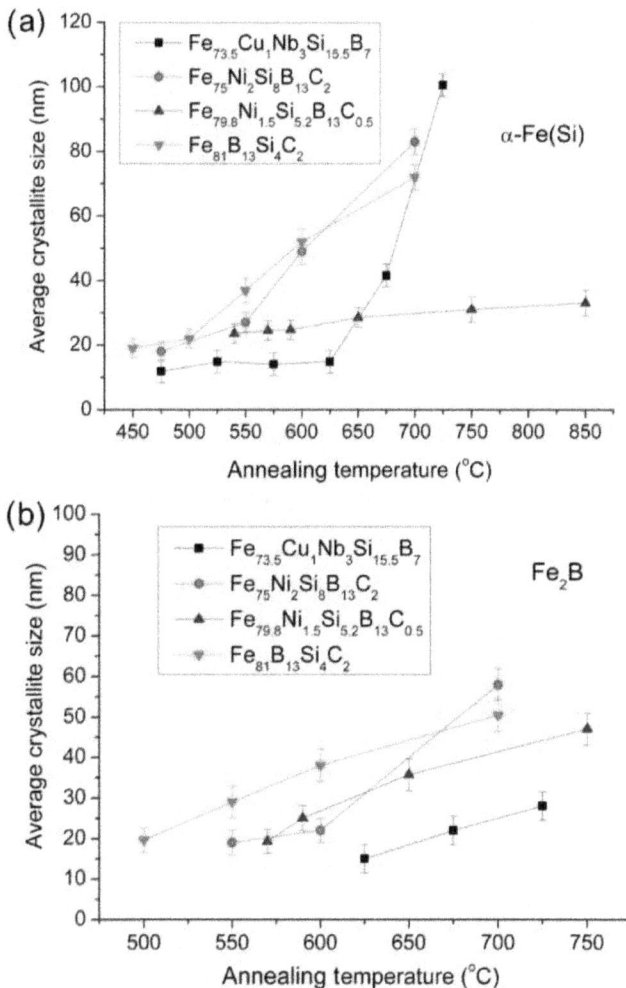

Figure 5.
Average crystallite size of the α-Fe(Si) (a) and Fe₂B (b) phases in the alloys containing 73–81 atomic % of iron, after thermal treatment at different temperatures.

maximal microhardness values correspond to the optimal microstructure, consisting of a composite involving nanocrystals embedded in amorphous matrix. This structure has a lower interfacial energy than purely amorphous or purely crystalline structure with crystal/crystal interface, suppressing propagation of shear bands and cracks along the interfaces [20, 26, 30]. At higher temperatures of thermal treatment, the dominant crystal/crystal interface with higher interfacial energy leads to easier shear band and crack propagation, yielding lower microhardness values.

3.4.2 Thermomagnetic measurements

Thermomagnetic measurements on heating [18, 20, 25, 28] revealed thermally induced microstructural changes, influencing the magnetic properties of the alloys. All the studied alloys exhibit two Curie temperatures (**Figure 7**), one corresponding to the as-prepared alloy (T_{c1}) and the second one corresponding to the Curie temperature of the alloy in the crystallized form (T_{c2}) (**Table 1**). The alloys

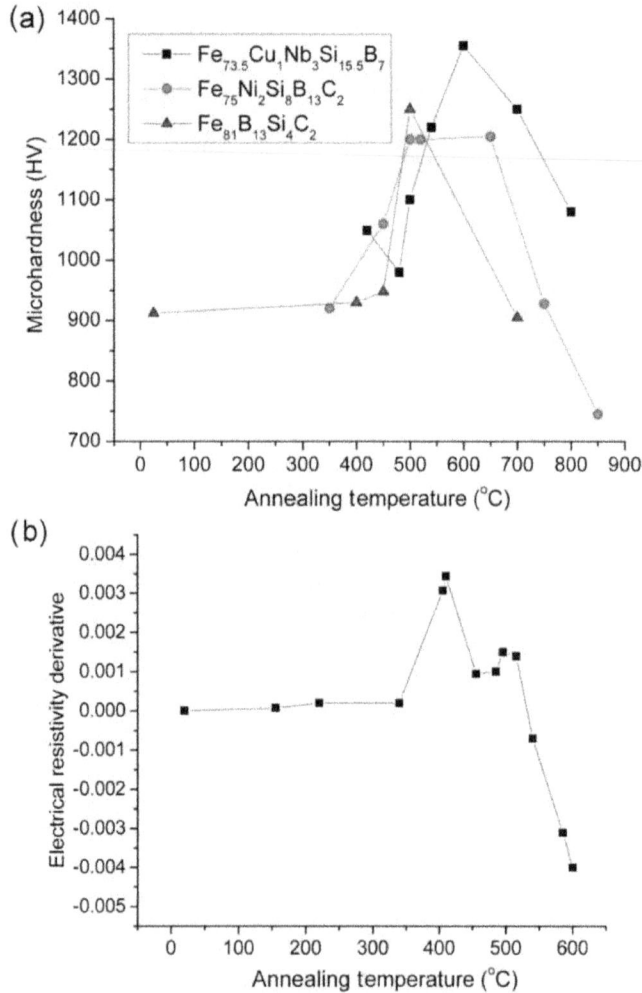

Figure 6.
Microhardness values of the $Fe_{73.5}Cu_1Nb_3Si_{15.5}B_7$, $Fe_{75}Ni_2Si_8B_{13}C_2$, and $Fe_{81}B_{13}Si_4C_2$ alloys after annealing at different temperatures (a) and the first derivative of the curve of temperature dependence of electrical resistivity for $Fe_{81}B_{13}Si_4C_2$ alloys (reprinted from ref. [29] with permission of Elsevier) (b).

$Fe_{75}Ni_2Si_8B_{13}C_2$ and $Fe_{81}B_{13}Si_4C_2$ exhibit similar values of the first Curie temperature, as a result of similarities in their chemical composition including high Fe content, and equal percentages of B and C. The lowest values of the first Curie temperature are observed for the $Fe_{73.5}Cu_1Nb_3Si_{15.5}B_7$ and $Fe_{40}Ni_{40}P_{14}B_6$ alloys. In the case of $Fe_{73.5}Cu_1Nb_3Si_{15.5}B_7$ alloy, low value of the first Curie temperature is provoked by the presence of Nb. It is well known that the addition of Nb reduces the Curie temperature of the amorphous phase by around 25% per atomic percent of Nb, while the influence of Cu is negligible [39]. Relatively low Fe content, relatively high Ni content, and the presence of P in the amorphous $Fe_{40}Ni_{40}P_{14}B_6$ alloy result in low value of the Curie temperature of this alloy. This is a consequence of the facts that Ni has lower Curie temperature and lower magnetic moment than Fe and the P addition has a decreasing effect on magnetic moment [40].

The beginning of crystallization process (**Figure 7**), as a result of formation of various magnetic crystalline phases, leads to an increase in magnetic moment of polycrystalline alloys. The manner of magnetic moment growth during the crystallization and subsequent decline when approaching the T_{c2} are determined by phase compositions of individual crystallized alloys. Thus, for example, in the case of $Fe_{73.5}Cu_1Nb_3Si_{15.5}B_7$, a rise in magnetic moment can be observed up to around 550°C, and then its drop starts, moving toward the Curie temperature of the formed crystalline phases. It should be noted that for the FINEMET-type alloys, to which $Fe_{73.5}Cu_1Nb_3Si_{15.5}B_7$ belongs, literature data [3, 41] usually include only the second value of Curie temperature because of its importance for practical application, since these alloys are mostly used in nanocrystalline form obtained by partial crystallization of amorphous precursor. Similarity of the T_{c2} values of the $Fe_{75}Ni_2Si_8B_{13}C_2$ and $Fe_{81}B_{13}Si_4C_2$ alloys results from their very similar phase composition in the fully crystalline form. However, the lowest T_{c2} value was observed for the $Fe_{40}Ni_{40}P_{14}B_6$ alloy, because the phases γ-(Fe,Ni) and $(Fe,Ni)_3(P,B)$ which constitute fully crystalline alloy are characterized by lower Curie temperature values than the α-Fe(Si) and Fe_2B phases forming the alloys with Fe as the dominant component.

3.4.3 Electrical resistivity measurements

Electrical resistivity measurements performed on the alloys containing 73–81 atomic % of iron [15, 27, 29, 42], at room temperature, reveal that the as-prepared $Fe_{79.8}Ni_{1.5}Si_{5.2}B_{13}C_{0.5}$ and $Fe_{81}B_{13}Si_4C_2$ alloys exhibit slightly lower electrical resistivity values, and better electronic conductivity, than the $Fe_{73.5}Cu_1Nb_3Si_{15.5}B_7$ and $Fe_{75}Ni_2Si_8B_{13}C_2$ alloys (**Table 3**), which is attributed to their somewhat higher iron content. As expected, after heating at different temperatures, each structural transformation is followed by certain changes in the trend of temperature dependence of electrical resistivity [15, 27, 29, 42].

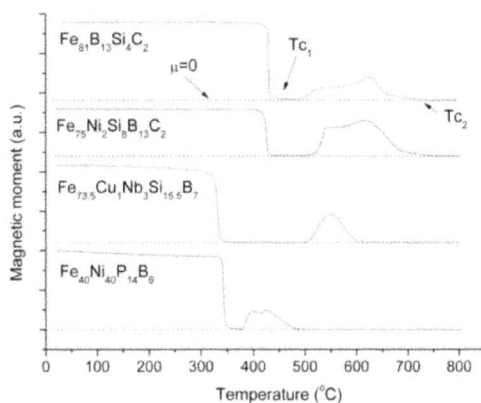

Figure 7.
Thermomagnetic curves recorded at 4°C/min.

Alloy	Electrical resistivity (μΩm)
$Fe_{81}B_{13}Si_4C_2$	1.71
$Fe_{79.8}Ni_{1.5}Si_{5.2}B_{13}C_{0.5}$	1.73
$Fe_{75}Ni_2Si_8B_{13}C_2$	2.27
$Fe_{73.5}Cu_1Nb_3Si_{15.5}B_7$	2.13

Table 3.
Electrical resistivity of the as-prepared alloys containing 73–81 atomic % of iron at room temperature.

The influence of thermally induced structural transformations on electrical resistivity of amorphous alloy can be illustrated with the example of $Fe_{75}Ni_2Si_8B_{13}C_2$ alloys [27]. In the temperature range 20–500°C, thermal treatment causes an increase in electrical resistivity (**Table 4**), where the slightly faster growth in the region 250–400°C corresponds to the structural relaxation, while the sharp increase occurs near the Curie point (400–430°C) [27]. Crystallization process, which starts at around 500°C, involves the sudden decline in electrical resistivity, since the ordered structure possesses lower electrical resistivity than the amorphous one. The second heating of the crystallized alloy results in linear growth of electrical resistivity with temperature [27], which is typical behavior of electronic (metal) conductors.

Measurement of electrical resistivity of the $Fe_{81}B_{13}Si_4C_2$ alloy after thermal treatment represents a good example of the situation when the functional properties are more sensitive to microstructural changes than thermal analysis. Derivative curve of the temperature dependence of electrical resistivity exhibits two well-defined maxima in the crystallization region (**Figure 6b**) [29], indicating that the crystallization in this case is a multistep process, although it occurs as a single peak in the DSC curve.

3.5 Crystallization kinetics

The knowledge of crystallization kinetics, besides thermal stability, is very important for usage of these alloys in modern technology, in order to estimate their applicability. The increase in heating rate leads to a shift in DSC peak temperature toward the region of higher temperatures [19, 21–24], showing that the observed processes are thermally activated, allowing the application of Arrhenius equation for kinetic description of the examined processes.

The kinetics of single-step solid-state phase transformation can be described using the equation:

$$\beta \frac{d\alpha}{dT} = A \exp\left(\frac{-E_a}{RT}\right) f(\alpha) \tag{1}$$

where T is the temperature, R is the gas constant, α is the conversion degree, β is the heating rate, $f(\alpha)$ is the conversion function representing the kinetic model, E_a is the activation energy, and A is the pre-exponential factor. The two last mentioned parameters are Arrhenius parameters, while the set including E_a, A and $f(\alpha)$ represents the kinetic triplet. For full kinetic description of a process, determination of kinetic triplet is required. Practical significance of kinetic triplets is determination of material lifetimes related to structural stability of materials and process rates [43].

Most of the observed crystallization DSC peaks are asymmetric as a result of complexity of crystallization processes involving more than one crystallization step. In order to study the kinetics of individual steps, complex crystallization peak deconvolution by application of appropriate mathematical procedure [19, 21–24] is required. For confirmation of single-step processes, isoconversional methods [43–49] are used.

Crystallization apparent activation energies for the formation of individual phases in the examined amorphous alloys, determined using Kissinger method [44], are presented in **Table 5**. The values obtained for the α-Fe(Si) phase are in the range 300–400 kJ/mol, while 200–350 kJ/mol are those determined for the Fe_2B phase. For all the crystalline phases in all the alloys examined, relatively high E_a values are obtained, probably as a result of cooperative participation of a large number of atoms in each step of the transformations [36]. The E_a values, obtained

Temperature (°C)	Electrical resistivity ($\mu\Omega$m)
20	2.268
100	2.282
150	2.296
200	2.310
250	2.331
350	2.408
400	2.492
410	2.548
440	2.576
530	2.604
540	2.604
545	2.492
550	2.352

Table 4.
Electrical resistivity measurements performed on the $Fe_{75}Ni_2Si_8B_{13}C_2$ alloy after thermal treatment at different temperatures.

using various methods [19, 21–24], are in agreement with the literature overall E_a values corresponding to the similar systems [36, 50, 51].

The alloys containing 73–81 atomic % of Fe, except the $Fe_{81}B_{13}Si_4C_2$, have lower crystallization apparent activation energy for the Fe_2B phase than that of the α-Fe(Si) phase by approximately 25%. This is a consequence of the creation of favorable conditions for crystallization of Fe_2B phase by enrichment of amorphous matrix with B caused by crystallization of α-Fe(Si) grains. The similar values of apparent activation energy of crystallization for the α-Fe(Si) and Fe_2B phases in the $Fe_{81}B_{13}Si_4C_2$ alloy can be explained by the presence of crystalline phase in amount of around 5% in the as-prepared structure acting as crystallization seeds and facilitating the crystallization of the α-Fe(Si) phase from the amorphous matrix. Higher value of apparent activation energy of crystallization of α-Fe(Si) can be observed for the $Fe_{79.8}Ni_{1.5}Si_{5.2}B_{13}C_{0.5}$ alloy due to the high thermal stability of this alloy, which originates from its optimal chemical composition. In the case of the alloy with high Ni content, formation of the bcc structure entails somewhat higher apparent activation energy (**Table 5**).

Kinetic analysis [19, 21–24] reveals that the conditions for application of the JMA model, most commonly used for kinetic description of transformations that consisted of nucleation and crystal growth processes, are not entirely fulfilled for any crystallization step in the alloys examined. Actually, for all crystallization steps, the shape of the Málek's curves [52] corresponds to the JMA model, but the maxima of the $z(\alpha)$ functions are shifted toward lower α values. Nucleation, which does not occur only in the early stages of transformations, and hard impingement effects corresponding to anisotropic crystal growth are the main contributors to such behavior. Anisotropic crystal growth is also indicated by the appearance of preferential orientation, observed during microstructural analysis [17]. Considering good accordance among the Málek's curves obtained at different heating rates, it can be concluded that the mechanism of the studied process does not change with heating rate in the range of heating rates examined. Autocatalytic Šesták-Berggren model, in two-parameter form $f(\alpha) = \alpha^M(1-\alpha)^N$, best describes the kinetics of crystallization, for all crystallization steps [19, 21–24]. Conversion functions of individual crystallization steps, in different alloys, are presented in **Table 5**. By introducing the kinetic triplets of individual crystallization steps into the equation for the solid-state transformation rate, with corresponding normalization and summation, simulated DSC curves can be obtained, which are, for the studied processes, in full accordance with experimental DSC curves [19, 21, 23], confirming the validity of the obtained kinetic triplets (**Figure 8**).

More information on crystallization mechanism can be obtained by considering values of local Avrami exponent, n [53]. Local Avrami exponent as well as the manner of its change with the progress of the process can indicate a certain transformation mechanism. For all crystallization steps of the examined alloys, decline in n value with the progress of transformation is observed (**Figure 9**) [19, 21]. This suggests the occurrence of impingement during the crystal growth, which was also indicated by microstructural analysis, as mentioned previously [19–21]. For non-isothermal measurements, at constant heating rates, conversion degree which corresponds to the position of the transformation rate maximum (α_p) suggests the anisotropic crystal growth as the prevailing type of impingement [54]. This includes blocking effects of growing particles occurring earlier than those for the isotropic growth, leading to hard impingement and to deviation from the classical JMA model [54]. Anisotropic crystal growth was also suggested by the existence of preferential orientation [17].

After determining the kinetic triplets, the lifetime of the alloys against crystallization which reflects their thermal stability as well as the stability of their functional properties is estimated. For the conversion degree of 5%, at room temperature, the alloys exhibit high lifetime values (10^{27}–10^{39} years) (**Table 6**), indicating that these materials are very stable at room temperature, in spite of their thermodynamic and kinetic metastability [21, 23]. Nevertheless, an increase in the

Phase	Alloy	E_a (kJ mol^{-1})	lnA (A/min^{-1})	$f(\alpha)$
α-Fe(Si)	Fe$_{81}$B$_{13}$Si$_4$C$_2$	320 ± 10	48 ± 2	$\alpha^{0.69}(1-\alpha)^{0.99}$
	Fe$_{79.8}$Ni$_{1.5}$Si$_{5.2}$B$_{13}$C$_{0.5}$	399 ± 6	58 ± 2	$\alpha^{0.98}(1-\alpha)^{1.20}$
	Fe$_{75}$Ni$_2$Si$_8$B$_{13}$C$_2$	298 ± 7	44 ± 1	$\alpha^{0.51}(1-\alpha)^{1.16}$
	Fe$_{73.5}$Cu$_1$Nb$_3$Si$_{15.5}$B$_7$	335 ± 7	49 ± 1	$\alpha^{0.46}(1-\alpha)^{1.20}$
Fe$_3$B	Fe$_{81}$B$_{13}$Si$_4$C$_2$	332 ± 5	50 ± 1	$\alpha^{0.69}(1-\alpha)^{0.93}$
	Fe$_{75}$Ni$_2$Si$_8$B$_{13}$C$_2$	230 ± 10	33 ± 3	$\alpha^{0.64}(1-\alpha)$
Fe$_2$B	Fe$_{81}$B$_{13}$Si$_4$C$_2$	340 ± 20	50 ± 3	$\alpha^{0.78}(1-\alpha)^{0.92}$
	Fe$_{79.8}$Ni$_{1.5}$Si$_{5.2}$B$_{13}$C$_{0.5}$	300 ± 10	43 ± 2	$\alpha(1-\alpha)^{1.30}$
	Fe$_{75}$Ni$_2$Si$_8$B$_{13}$C$_2$	210 ± 20	29 ± 4	$\alpha^{0.62}(1-\alpha)$
	Fe$_{73.5}$Cu$_1$Nb$_3$Si$_{15.5}$B$_7$	260 ± 20	37 ± 3	$\alpha^{0.51}(1-\alpha)^{1.30}$
Fe$_{16}$Nb$_6$Si$_7$	Fe$_{73.5}$Cu$_1$Nb$_3$Si$_{15.5}$B$_7$	490 ± 10	60 ± 2	$\alpha(1-\alpha)^{1.40}$
Fe$_2$Si	Fe$_{73.5}$Cu$_1$Nb$_3$Si$_{15.5}$B$_7$	470 ± 30	58 ± 5	$\alpha^{0.60}(1-\alpha)^{1.10}$
α-(Fe,Ni)	Fe$_{40}$Ni$_{40}$P$_{14}$B$_6$	450 ± 20	82 ± 3	$\alpha^{0.53}(1-\alpha)^{1.11}$
γ-(Fe,Ni)	Fe$_{40}$Ni$_{40}$P$_{14}$B$_6$	450 ± 30	80 ± 5	$\alpha^{0.50}(1-\alpha)^{1.15}$
(Fe,Ni)$_3$(P,B)	Fe$_{40}$Ni$_{40}$P$_{14}$B$_6$	460 ± 30	81 ± 6	$\alpha^{0.48}(1-\alpha)^{1.18}$

Table 5.
Kinetic triplets of individual crystallization steps determined for different alloys.

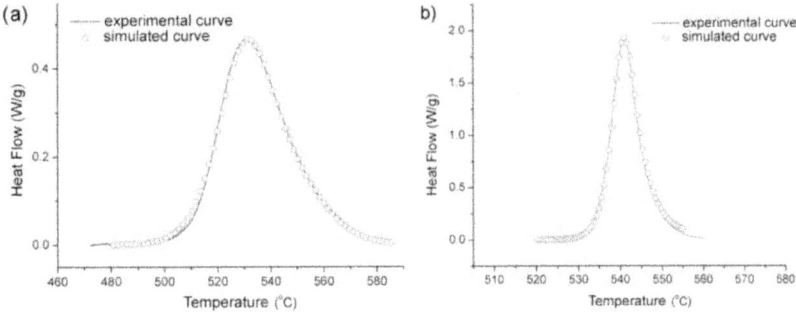

Figure 8.
Examples of comparison of experimental DSC curves at 8°C/min and the curves simulated with determined kinetic triplets of individual crystallization steps: $Fe_{73.5}Cu_1Nb_3Si_{15.5}B_7$ alloy, peak 1 (a), and $Fe_{79.8}Ni_{1.5}Si_{5.2}B_{13}C_{0.5}$ alloy (b).

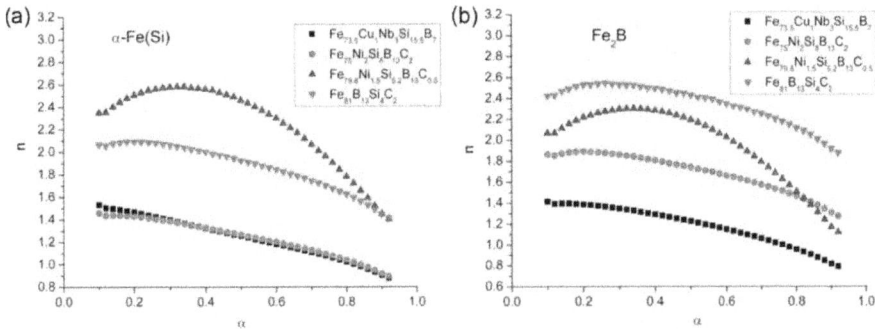

Figure 9.
Local values of Avrami exponent of α-Fe(Si) (a) and Fe_2B (b) phases in different alloys at 5 °C/min.

Alloy	Lifetime (year)
$Fe_{81}B_{13}Si_4C_2$	2.2×10^{29}
$Fe_{79.8}Ni_{1.5}Si_{5.2}B_{13}C_{0.5}$	3.6×10^{39}
$Fe_{75}Ni_2Si_8B_{13}C_2$	2.5×10^{27}
$Fe_{73.5}Cu_1Nb_3Si_{15.5}B_7$	2.2×10^{30}
$Fe_{40}Ni_{40}P_{14}B_6$	3.3×10^{38}

Table 6.
Estimated values of the lifetime of the alloys against crystallization at room temperature, determined for conversion degree of 5%.

temperature of thermal treatment leads to an exponential decline in the values of estimated lifetime against crystallization, which amounts to only several minutes at the temperature of the onset of crystallization process [21, 23]. At room temperature, the amorphous $Fe_{79.8}Ni_{1.5}Si_{5.2}B_{13}C_{0.5}$ alloy shows lifetime value by several orders of magnitude higher than those of the other alloys containing 73–81 atomic % of Fe, which is in accordance with its higher thermal stability. In spite of crystallizing at lower temperatures than the alloys with 73–81 atomic % of Fe, the alloy containing 40 atomic % of Fe shows higher thermal stability at room temperature, manifested by higher lifetime values than those of the alloys containing Fe as the dominant component (**Table 6**).

4. Conclusion

A detailed study of five iron-based amorphous alloys with the compositions $Fe_{81}Si_4B_{13}C_2$, $Fe_{79.8}Ni_{1.5}Si_{5.2}B_{13}C_{0.5}$, $Fe_{75}Ni_2Si_8B_{13}C_2$, $Fe_{73.5}Cu_1Nb_3Si_{15.5}B_7$, and $Fe_{40}Ni_{40}P_{14}B_6$ revealed that the alloy behavior in terms of mechanism, thermodynamics, and kinetics of thermally induced microstructural transformations, as well as the functional properties, is significantly influenced by chemical composition of the alloy. The highest thermal stability among the studied alloys was observed for the $Fe_{79.8}Ni_{1.5}Si_{5.2}B_{13}C_{0.5}$ alloy due to its optimal chemical composition. Crystallization changed alloy microstructure and morphology, making the alloys grainy and more porous, influencing the functional properties of the alloys. Crystalline α-Fe(Si) and Fe_2B phases were observed to be the final crystallization products in all the alloys with Fe as the dominant component. Kinetic analysis of individual crystallization steps, performed after peak deconvolution, revealed around 25% lower apparent activation energy values of the Fe_2B phase than those of the α-Fe(Si) phase, for most of the studied alloys, as a result of promoted Fe_2B crystallization by formation of α-Fe(Si) grains and an enrichment of the amorphous matrix with boron. Relatively high stability of the studied alloys against crystallization was observed at room temperature in spite of thermodynamic metastability and kinetic metastability of amorphous materials, with its abrupt drop at increased temperatures.

Acknowledgements

This research was supported by the Ministry of Education, Science and Technological Development of the Republic of Serbia, under the Project No. OI172015.

Author details

Milica M. Vasić[1], Dušan M. Minić[2] and Dragica M. Minić[1*]

1 Faculty of Physical Chemistry, University of Belgrade, Belgrade, Serbia

2 Military Technical Institute, Belgrade, Serbia

*Address all correspondence to: dminic@ffh.bg.ac.rs

IntechOpen

References

[1] Takayama S. Amorphous structures and their formation and stability. Journal of Materials Science. 1976;**11**:164-185

[2] Flohrer S, Herzer G. Random and uniform anisotropy in soft magnetic nanocrystalline alloys. Journal of Magnetism and Magnetic Materials. 2010;**322**:1511-1514

[3] Suryanarayana C, Inoue A. Iron-based bulk metallic glasses. International Materials Review. 2013;**58**:131-166

[4] Shen B, Inoue A. Superhigh strength and good soft-magnetic properties of (Fe,Co)-B-Si-Nb bulk glassy alloys with high glass-forming ability. Applied Physics Letters. 2004;**85**:4911-4913

[5] Pang S, Zhang T, Asami K, Inoue A. Effects of chromium on the glass formation and corrosion behavior of bulk glassy Fe-Cr-Mo-C-B alloys. Materials Transactions. 2002;**43**:2137-2142

[6] Gavrilović A, Rafailović LD, Minić DM, Wosik J, Angerer P, Minić DM. Influence of thermal treatment on structure development and mechanical properties of amorphous $Fe_{73.5}Cu_1Nb_3Si_{15.5}B_7$ ribbon. Journal of Alloys and Compounds. 2011;**509S**:S119-S122

[7] McHenry ME, Willard MA, Laughlin DE. Amorphous and nanocrystalline materials for applications as soft magnets. Progress in Materials Science. 1999;**44**:291-433

[8] Gu XJ, Poon SJ. Mechanical properties of iron-based bulk metallic glasses. Journal of Materials Research. 2007;**22**:344-351

[9] Inoue A, Shen B. Formation and soft magnetic properties of Fe-B-Si-Zr bulk glassy alloys with high saturation magnetization above 1.5 T. Materials Transactions. 2002;**43**:2350-2353

[10] Inoue A. High strength bulk amorphous alloys with low critical cooling rates. Materials Transactions. 1995;**36**:866-875

[11] Greer AL. Confusion by design. Nature. 1993;**366**:303-304

[12] Du SW, Ramanujan RV. Crystallization and magnetic properties of $Fe_{40}Ni_{38}B_{18}Mo_4$ amorphous alloy. Journal of Non-Crystalline Solids. 2005;**351**:3105-3113

[13] Gleiter H. Nanocrystalline materials. Progress in Materials Science. 1989;**33**:223-315

[14] Kulik T. Nanocrystallization of metallic glasses. Journal of Non-Crystalline Solids. 2001;**287**:145-161

[15] Blagojević VA, Minić DM, Vasić M, Minić DM. Thermally induced structural transformations and their effect on functional properties of Fe $_{89.8}Ni_{1.5}Si_{5.2}B_3C_{0.5}$ amorphous alloy. Materials Chemistry and Physics. 2013;**142**:207-212

[16] Vasić MM, Minić DM, Blagojević VA, Žák T, Pizúrová N, David B, et al. Thermal stability and mechanism of thermally induced crystallization of $Fe_{73.5}Cu_1Nb_3Si_{15.5}B_7$ amorphous alloy. Acta Physica Polonica, A. 2015;**128**:657-660

[17] Blagojević VA, Vasić M, David B, Minić DM, Pizúrová N, Žák T, et al. Thermally induced crystallization of $Fe_{73.5}Cu_1Nb_3Si_{15.5}B_7$ amorphous alloy. Intermetallics. 2014;**45**:53-59

[18] Vasić MM, Roupcová P, Pizúrová N, Stevanović S, Blagojević VA, Žák T, et al. Thermally induced structural

transformations of $Fe_{40}Ni_{40}P_{14}B_6$ amorphous alloy. Metallurgical and Materials Transactions A. 2016;**47A**:260-267

[19] Vasić MM, Blagojević VA, Begović NN, Žák T, Pavlović VB, Minić DM. Thermally induced crystallization of amorphous $Fe_{40}Ni_{40}P_{14}B_6$ alloy. Thermochimica Acta. 2015;**614**:129-136

[20] Blagojević VA, Vasić MM, David B, Minić DM, Pizúrová N, Žák T, et al. Microstructure and functional properties of $Fe_{73.5}Cu_1Nb_3Si_{15.5}B_7$ amorphous alloy. Materials Chemistry and Physics. 2014;**145**:12-17

[21] Vasić MM, Minić DM, Blagojević VA, Minić DM. Kinetics and mechanism of thermally induced crystallization of amorphous $Fe_{73.5}Cu_1Nb_3Si_{15.5}B_7$ alloy. Thermochimica Acta. 2014;**584**:1-7

[22] Blagojević VA, Vasić M, Minić DM, Minić DM. Kinetics and thermodynamics of thermally induced structural transformations of amorphous $Fe_{75}Ni_2Si_8B_{13}C_2$ alloy. Thermochimica Acta. 2012;**549**:35-41

[23] Vasić M, Minić DM, Blagojević VA, Minić DM. Mechanism and kinetics of crystallization of amorphous $Fe_{81}B_{13}Si_4C_2$ alloy. Thermochimica Acta. 2013;**572**:45-50

[24] Vasić M, Minić DM, Blagojević VA, Minić DM. Mechanism of thermal stabilization of $Fe_{89.8}Ni_{1.5}Si_{5.2}B_3C_{0.5}$ amorphous alloy. Thermochimica Acta. 2013;**562**:35-41

[25] Minić DM, Blagojević VA, Maričić AM, Žák T, Minić DM. Influence of structural transformations on functional properties of $Fe_{75}Ni_2Si_8B_{13}C_2$ amorphous alloy. Materials Chemistry and Physics. 2012;**134**:111-115

[26] Blagojević VA, Minić DM, Žák T, Minić DM. Influence of

thermal treatment on structure and microhardness of $Fe_{75}Ni_2Si_8B_{13}C_2$ amorphous alloy. Intermetallics. 2011;**19**:1780-1785

[27] Minić DM, Maričić AM. Influence of heating on electric and magnetic properties of $Fe_{75}Ni_2B_{13}Si_8C_2$ amorphous alloy. Materials Science and Engineering B. 2010;**172**:127-131

[28] Minić DM, Minić DM, Žák T, Roupcová P, David B. Structural transformations of $Fe_{81}B_{13}Si_4C_2$ amorphous alloy induced by heating. Journal of Magnetism and Magnetic Materials. 2011;**323**:400-404

[29] Minić DM, Minić DG, Maričić A. Stability and crystallization of $Fe_{81}B_{13}Si_4C_2$ amorphous alloy. Journal of Non-Crystalline Solids. 2009;**355**:2503-2507

[30] Minić DM, Blagojević VA, Minić DM, Gavrilović A, Rafailović L, Žák T. Influence of microstructure on microhardness of $Fe_{81}Si_4B_{13}C_2$ amorphous alloy after thermal treatment. Metallurgical and Materials Transactions A. 2011;**42A**:4106-4112

[31] Inorganic Crystals Structure Database (ICSD). Eggenstein-Leopoldshafen, Germany: FIZ Karlsruhe; 2014

[32] JCPDS PDF-2 Database. Newton Square, PA, USA: ICDD; 2005

[33] Crystallography Open Database. Available from: http://www. crystallography.net [Accessed: September 1, 2014, June 15, 2016]

[34] Lutterotti L. Total pattern fitting for the combined size-strain-stress-texture determination in thin film diffraction. Nuclear Instruments and Methods in Physics Research-Section B. 2010;**268**:334-340

[35] Scherrer P. Göttinger Nachrichten. Journal of Mathematical Physics. 1918;**2**:98-100

[36] Kaloshkin SD, Tomilin IA. The crystallization kinetics of amorphous alloys. Thermochimica Acta. 1996;**280/281**:303-317

[37] Qin J, Gu T, Yang L, Bian X. Study on the structural relationship between the liquid and amorphous $Fe_{78}Si_9B_{13}$ alloys by ab initio molecular dynamics simulation. Applied Physics Letters. 2007;**90**:201909

[38] Ma H, Wang W, Zhang J, Li G, Cao C, Zhang H. Crystallization and corrosion resistance of $(Fe_{0.78}Si_{0.09}B_{0.13})_{100-x}Ni_x$ (x=0, 2 and 5) glassy alloys. Journal of Materials Science and Technology. 2011;**27**:1169-1177

[39] Miguel C, Kaloshkin S, Gonzalez J, Zhukov A. Curie temperature behaviour on annealing of Finemet type amorphous alloys. Journal of Non-Crystalline Solids. 2003;**329**:63-66

[40] Becker JJ, Luborsky FE, Walter JL. Magnetic moments and curie temperatures of $(Fe, Ni)_{80}(P, B)20$ amorphous alloys. IEEE Transactions on Magnetics. 1977;**13**:988-991

[41] Yoshizawa Y, Oguma S, Yamauchi K. New Fe-based soft magnetic alloys composed of ultrafine grain structure. Journal of Applied Physics. 1988;**64**:6044-6046

[42] Maričić AM, Minić DM, Blagojević VA, Kalezić-Glišović A, Minić DM. Effect of structural transformations preceding crystallization on functional properties of $Fe_{73.5}Cu_1Nb_3Si_{15.5}B_7$ amorphous alloy. Intermetallics. 2012;**21**:45-49

[43] Vyazovkin S, Burnham AK, Criado JM, Pérez-Maqueda LA, Popescu C, Sbirrazzuoli N. ICTAC kinetics committee recommendations for performing kinetic computations on thermal analysis data. Thermochimica Acta. 2011;**520**:1-19

[44] Kissinger HE. Reaction kinetics in differential thermal analysis. Analytical Chemistry. 1957;**29**:1702-1706

[45] Ortega A. A simple and precise linear integral method for isoconversional data. Thermochimica Acta. 2008;**474**:81-86

[46] Ozawa T. A new method of analyzing thermogravimetric data. Bulletin of the Chemical Society of Japan. 1965;**38**:1881-1886

[47] Flynn JH, Wall LA. A quick, direct method for the determination of activation energy from thermogravimetric data. Journal of Polymer Science Part C: Polymer Letters. 1966;**4**:323-328

[48] Vyazovkin S. Modification of the integral isoconversional method to account for variation in the activation energy. Journal of Computational Chemistry. 2001;**22**:178-183

[49] Vyazovkin S. Evaluation of activation energy of thermally stimulated solid-state reactions under arbitrary variation of temperature. Journal of Computational Chemistry. 1997;**18**:393-402

[50] Wang Y, Xu K, Li Q. Comparative study of non-isothermal crystallization kinetics between $Fe_{80}P_{13}C_7$ bulk metallic glass and melt-spun glassy ribbon. Journal of Alloys and Compounds. 2012;**540**:6-15

[51] Santos DS, Santos DR. Crystallization kinetics of Fe-B-Si metallic glasses. Journal of Non-Crystalline Solids. 2002;**304**:56-63

[52] Málek J. Kinetic analysis of crystallization processes in amorphous

materials. Thermochimica Acta. 2000;**355**:239-253

[53] Blazquez JS, Conde CF, Conde A. Non-isothermal approach to isokinetic crystallization processes: Application to the nanocrystallization of HITPERM alloys. Acta Materialia. 2005;**53**:2305-2311

[54] Liu F, Song SJ, Sommer F, Mittemeijer EJ. Evaluation of the maximum transformation rate for analyzing solid-state phase transformation kinetics. Acta Materialia. 2009;**57**:6176-6190

Chapter 4

Phase Separation in Ce-Based Metallic Glasses

Dharmendra Singh, Kiran Mor, Devinder Singh and Radhey Shyam Tiwari

Abstract

In this chapter, the results of our recent studies on the role of Ga substitution in place of Al in $Ce_{75}Al_{25-x}Ga_x$ (x = 0, 0.01, 0.1, 0.5, 1, 2, 4, and 6) metallic glasses (MGs) have been discussed with the aim to understand the genesis of phase separation. X-ray diffraction (XRD) study reveals two broad diffuse peaks corresponding to the coexistence of two amorphous phases. In order to see any change in the behavior of $4f$ electron of Ce, X-ray absorption spectroscopy (XAS) has been carried out for $Ce_{75}Al_{25-x}Ga_x$ MGs. From the XAS results, it is evident that for x = 0, the spectrum exhibits only a $4f^1$ component, which basically shows a pure localized configuration of electron. After the addition of Ga, $4f$ electrons of Ce atoms denoted by $4f^0$ are getting delocalized. Thus, the phase separation in $Ce_{75}Al_{25-x}Ga_x$ is taking place, owing to the formation of two types of amorphous phases having localized and delocalized $4f$ electrons of Ce atoms, respectively. It has been discussed how change in the electronic structure of Ce atoms may lead to phase separation in $Ce_{75}Al_{25-x}Ga_x$ alloys. Extensive TEM investigations have been done to study the phase separation in these alloys. The microstructural features have been compared with those obtained by phase field modeling.

Keywords: metallic glass, phase separation, X-ray absorption spectroscopy, transmission electron microscopy, phase field modeling

1. Introduction

In the past decades, considerable research attention has been given to rare-earth (RE)-based metallic glasses (MGs) due to their novel physical properties such as glass-forming ability [1] and mechanical [2, 3], magnetic [4], superplastic [5], and thermoplastic properties [6]. Thus, these MGs hold potential in many applications in the future. Many novel RE-based MGs, e.g., Ce-, La-, Y-, Er-, and Sm-based MGs, have been synthesized [7]. Among RE-based MGs, Ce-based MGs are of special interest due to their unusual behavior linked to $4f$ electrons [8]. Ce is the most abundant RE metal on earth. It is also one of the most reactive RE metal and oxidizes very readily even at room temperature. One of the key features of Ce is its variable valance states and electronic structure [9–11]. Thus to change the relative occupancy of the electronic levels, only a small amount of energy is required, e.g., a volume change of approximately 10% results when Ce is subjected to high pressure or low temperatures [9, 11]. Therefore, Ce-based MGs may possess structural and physical properties which are different from other known MGs [12].

IntechOpen

Recently, a pressure-induced devitrification behavior of $Ce_{75}Al_{25}$ MG ribbon has been reported [13–15]. Prior to our study, only few studies have been done on the substitution and mechanical behavior of $Ce_{75}Al_{25}$ glassy alloy [1, 16].

Any approach to the description of the amorphous structure suggests that it is a homogeneous isotropic structure. In fact, it turned out that the structure of amorphous phase in alloys cannot always be uniform and isotropic. One situation occurs in the case when the amorphous phase contains two or more metals with comparable scattering amplitude. In such systems, the appearance of inhomogeneity areas or two types of amorphous phases is much more pronounced, since the formation of regions with different chemical compositions leads to the appearance of at least two types of shortest distances between atoms, which naturally results in the phase separation and also affects various properties. The first report by Chen and Turnbull [17] on phase separation in Pd-Au-Si alloy has attracted considerable attention due to their unique microstructural variation of amorphous phases at different length scales. Following this, the possibility of phase separation in MG compositions has been investigated by many authors [18–20]. However, such a phase separation is incompatible with the glass-forming criteria of negative heat of mixing [21]. The models of MGs based on the nature of geometrical clusters [22] may be helpful in comprehending phase separation in these alloys. According to this model, the MGs have geometry incompatibility in main clusters with long-range translational orders and are joined by the cementing cluster known as glue cluster [23–33]. Sohn et al. reported two general schemes for the design of phase-separating MGs [34]. The first scheme refers to the selection of atom pairs having positive enthalpy of mixing, and the second one refers to the selection of additional alloying element which can enhance glass-forming ability. In the case of ternary- and higher-component alloys, the opposite nature of enthalpy of mixing between the pairs of binaries is possible. In MG systems phase separation will be due to the complex interplay of positive and negative enthalpies of mixing, e.g., in Gd-Zr-Al-Ni Mg alloy system, the enthalpy of mixing is positive for Gd-Zr atom pairs, and other pairs consist of negative enthalpy of mixing [34]. That's why phase separation is shown by MG system in amorphous state. Phase separation is exhibited by many alloy systems such as La-Zr-Al-Cu-Ni [35], Zr-Ti-Ni-Cu-Be [36], Zr-Gd-Co-Al [37], Cu-(Zr,Hf)-(Gd,Y)-Al [38], Cu-Zr-Al-Nb [39], and Gd-Hf-Co-Al [40]. However, there are very few ternary systems reported in literature which show phase separation. Wu et al. have studied ternary Pd-Ni-P alloy system and observed phase separation through spinodal decomposition [41]. It is worthwhile to mention here that so far no report is available prior to our present study where very sparse atomic percent (~ 0.01 at.%) addition of an element leads to phase separation in a binary system.

In this chapter, we present extensive investigations of amorphous phase formation in $Ce_{75}Al_{25-x}Ga_x$ alloys with a wide range of concentration of Ga (x = 0, 0.01, 0.1, 0.5, 1, 2, 4, and 6). Both Al and Ga are having the same valency (+3), comparable atomic radii (Ga, 1.41 Å; Al, 1.43 Å), and lying in the same group of the periodic table. Thus, the substitution of Al by Ga does not change the e/a ratio of Ce-Al alloy system (e/a = 1.39). It has been undertaken with a view to understanding the genesis of phase separation in this alloy system. The microstructural features arise due to phase separation which has been studied by transmission electron microscopy (TEM) and compared with those obtained by phase field modeling. The role of Ce electronic structure in phase separation has been discussed. It is important to mention that due to change in the electronic states of Ce, $4f$ electrons under high pressure, $Ce_{75}Al_{25}$ alloy undergoes polyamorphic transition [13, 42, 43]. One may expect that chemical pressure effect of Ga substitution in $Ce_{75}Al_{25}$ MG leads to change in the electronic structure of the Ce in this alloy [44]. Chemical pressure effect basically deals with the change in the electronic structure of atoms due to pressure, temperature, or alloying addition. Keeping these facts in view, extensive use of X-ray absorption spectroscopy (XAS) has been done to investigate $Ce_{75}Al_{25-x}Ga_x$ alloys. Our investigations have clearly demonstrated that two types of short range order (SRO) may set in $Ce_{75}Al_{25-x}Ga_x$ amorphous alloys [23]. This is due to delocalization of $4f$ electron with addition of Ga. The change in the electronic structure of Ce is considered as one of the important reasons for the phase separation in Ce-Al-Ga MG alloy system. The remarkable change in the behavior of glass transition with Ga substitution has been observed through DSC investigation [25–30]. The thermal stability of the studied materials has been discussed elsewhere, and for this we refer the readers to reference [27].

In this chapter, the effect of Ga substitution (with x as low as 0.01 at.%) on the phase separation has been discussed. The substitution of Ga at place of Al in various alloy systems has been extensively studied by our group [45–50]. The Ce-Al [51] and Ce-Ga [52] binaries have negative heat of mixing, while Ga-Al pair has very low positive heat of mixing, i.e., 0.7 KJ/mol [53]. It seems unlikely that the phase separation is caused by Ga-Al which has a very small positive heat of mixing. Hence, the alternative explanation for this has been called for. One may thus expect that the substitution of Ga on Al sites may lead to change in the electronic behavior of Ce $4f$ electrons (owing to chemical pressure effect) [54]. We have also discussed the effect of Ga substitution on the formation of nanoamorphous domains as well as on the nature of Ce $4f$ electronic states. It should be pointed out that pressure-induced delocalization of $4f$ electron (using XAS studies) has also been reported by other researchers [13, 42]. However, the partial delocalization of $4f$ electron of Ce atoms in $Ce_{75}Al_{25-x}Ga_x$ alloys due to Ga substitution has been pointed out for the first time based on XAS studies.

2. Materials and experimental procedure

The details of the preparation methods of $Ce_{75}Al_{25-x}Ga_x$ melt-spun alloys are reported elsewhere [2, 21]. The structural characterization has been carried out using X–ray diffractometer (X'Pert Pro PANalytical diffractometer) with CuK_α radiation. The electrolyte with 70% methanol and 30% nitric acid at 253 K has been used to thin the ribbons for TEM characterization. The TEM using FEI: Tecnai $20G^2$ electron microscope has been used to observe the thinned samples. Energy-dispersive X-ray analysis (EDX) attached to the TEM Tecnai 20 G^2 is obtained at 200 keV using 100 seconds exposure time and 4 µA beam current. The X-ray absorption spectroscopy (XAS) measurements on these samples at Ce L_3 edge were carried out in fluorescence mode with beamline (BL-9), INDUS-2 synchrotron source (2.5 GeV, 100 mA), at RRCAT, India.

3. Investigation of $Ce_{75}Al_{25-x}Ga_x$ (x = 0, 0.01, 0.1, 0.5, 1, 2, 4, and 6) alloys

3.1 A comparative X-ray diffraction investigation of $Ce_{75}Al_{25-x}Ga_x$ alloys

Figures 1 and **2** show the XRD patterns of $Ce_{75}Al_{25-x}Ga_x$ alloys at different Ga concentrations. For the alloy with x = 0, the broad halo peak is found within the angular range 28–35°. This indicates the formation of homogenous glassy phase in $Ce_{75}Al_{25}$ alloy. While for the alloys with x = 2–6, broad halo peak is found within the angular range 39–50°. The unusual effect was seen in the XRD pattern on substitution of 0.01 at.% Ga. The second diffuse peak with higher intensity can be seen at higher-angle side. With increase in the quantity of Ga (x = 0.1, 0.5, 1, 2, 4, and 6),

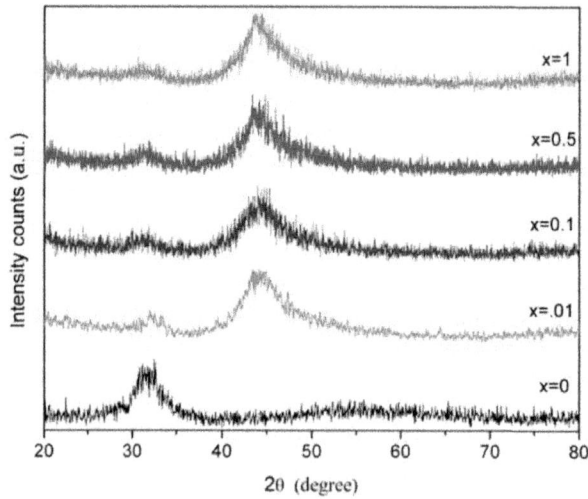

Figure 1.
XRD patterns of as-synthesized ribbons of $Ce_{75}Al_{25-x}Ga_x$ alloys (x = 0, 0.01, 0.1, 0.5, and 1) (reprinted with kind permission from Ref. [25], copyright 2016, Elsevier).

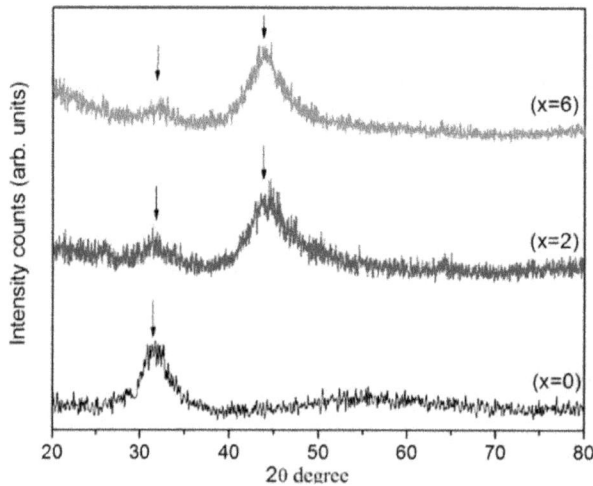

Figure 2.
XRD patterns of as-synthesized ribbons of $Ce_{75}Al_{25-x}Ga_x$ alloys (x = 0, 2, and 6) (reprinted with kind permission from Reference [27], copyright 2014, Elsevier).

the positions and intensities of the higher-angle diffuse peak remains almost the same for different concentrations of Ga. The formation of additional diffuse halo peak on the higher-angle side in the XRD pattern due to addition of such sparse amount of Ga refers to unusual effect.

The prominent low-angle peak (~32°) with low intensity has been observed for x = 0 with respect to Ga addition. The formation of two amorphous phases for the alloys with x = 0.01–6 has been depicted from the two diffuse peaks with different intensities in the XRD patterns of $Ce_{75}Al_{25-x}Ga_x$ alloys. It can be noticed that one hump is at its original position which indicates that the nature of short range order has not changed for pristine phase. The second diffuse peak appears at ~44° which indicates the significant change in the short range order. It may be pointed out that usually the hump in the XRD patterns for the large number of MGs occurs in the range of 26–38°. In the present case, the second hump is lying in the same range indicating that the SRO is very similar to the most common type of MGs. Similar observation of two humps has also been reported by Kim et al. for phase separation in $Ti_{45}Y_{11}Al_{24}Co_{20}$ metallic glass [54].

3.2 Comparative electron microscopic (TEM) investigation of $Ce_{75}Al_{25-x}Ga_x$ alloys

The TEM image of $Ce_{75}Al_{25}$ depicts homogenous contrast, and its corresponding selected area diffraction (SAD) shows single diffuse halo ring (c.f. **Figure 3(a)**). After Ga substitution, the presence of two different amorphous phases having two different contrasts can be seen in **Figure 3(b–f)**. There is one type of amorphous phase which is dispersed in the matrix of other amorphous phase. **Figure 3(b–f)** displays SAD patterns with two diffuse halos after Ga substitution. The analysis of domain size dispersed in the amorphous matrix has been carried out, and the domain size variation with Ga addition has been done using *IMAGE J* software. The value domain size (in nanometer) increases linearly with Ga addition and then obtains a saturation value, i.e., ~7 nm at x = 4 and beyond. In **Figure 3(b–f)**, insets

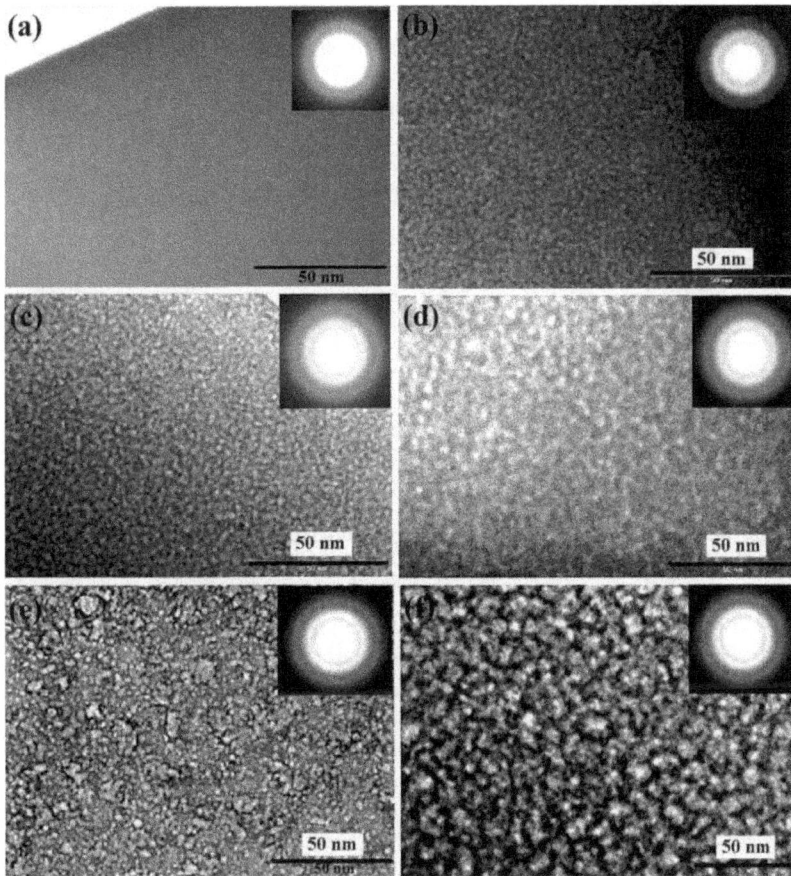

Figure 3.
Bright-field TEM microstructures and the corresponding selected area diffraction patterns (shown in inset) of $Ce_{75}Al_{25-x}Ga_x$ alloys with (a) x = 0, (b) x = 0.1, (c) x = 0.5, (d) x = 1, (e) x = 2, and (f) x = 4 (reprinted with kind permission from Reference [25], copyright 2016, Elsevier).

also show two diffuse halos from the matrix of one amorphous phase and dispersed (secondary) amorphous phase. The clear variation in the microstructure (**Figure 3**) due to Ga addition can be seen. However, in the XRD patterns, not much variation in the intensities of two humps is found. It can be said that the two humps are due to the presence of two types of "short range order" in coexisting amorphous phases.

3.3 Compositional analysis of $Ce_{75}Al_{25-x}Ga_x$ alloys through energy-dispersive X-ray analysis

The EDX spectra of $Ce_{75}Al_{25-x}Ga_x$ alloys (x = 0, 0.5, 1, and 4) are shown in **Figure 4(a–d)**. **Table 1** represents the average and nominal composition variations for the alloys with x = 0–6. The deviation reported is on the basis of measurements taken from four to six regions of the sample. The percentage experimental error in the case of Ga is found to be highest. The analysis shows Ga is responsible for contrast variation because of two kinds of amorphous domains in $Ce_{75}Al_{25-x}Ga_x$ alloys. Within the traceable limit of EDX, the presence of silicon (Si) could not be found. Because of very fine droplet-like features (<7 nm), it is not possible to characterize the variation of Ga in amorphous matrix as well as droplet-like structure. For compositional analysis in TEM, the probe size is ~50 μm at magnification of 13.5 k. That's why only nominal and average composition of Ga is shown.

3.4 X-ray absorption spectroscopy (XAS) investigation of $Ce_{75}Al_{25-x}Ga_x$ alloys

Figure 5 shows Ce L_3 edge XAS spectra as a function of addition of Ga in $Ce_{75}Al_{25}$ alloy. The spectrum exhibited by $Ce_{75}Al_{25}$ alloy is having only $4f^1$ component that gives a pure localized $4f^1$ configuration. It can be seen that in the XAS spectra of $Ce_{75}Al_{25}$, the signature of $4f^0$ electron is not present. The postedge feature

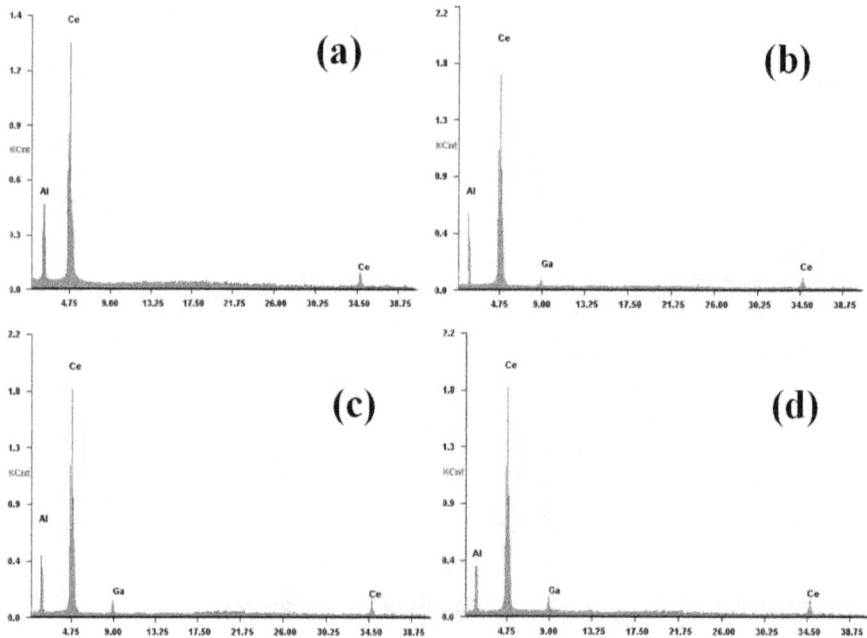

Figure 4.
Energy dispersive spectra of the melt-spun $Ce_{75}Al_{25-x}Ga_x$ alloys for (a) x = 0, (b) x = 0.5, (c) x = 1, and (d) x = 4 alloys (reprinted with kind permission from Reference [25], copyright 2016, Elsevier).

S. No.	x	Nominal composition	Average EDX composition*
1	0	$Ce_{75}Al_{25}$	$Ce_{74.8 \pm 1.5}Al_{25.0 \pm 0.8}$
2	0.1	$Ce_{75}Al_{24.9}Ga_{0.1}$	$Ce_{74.8 \pm 1.5}Al_{24.9 \pm 1.7}Ga_{0.1 \pm 0.1}$
3	0.5	$Ce_{75}Al_{24.5}Ga_{0.5}$	$Ce_{75.1 \pm 3.0}Al_{24.2 \pm 3.0}Ga_{0.7 \pm 0.3}$
4	1.0	$Ce_{75}Al_{24.0}Ga_{1.0}$	$Ce_{74.5 \pm 1.7}Al_{24.3 \pm 0.9}Ga_{1.2 \pm 0.9}$
5	2.0	$Ce_{75.0}Al_{23.0}Ga_{2.0}$	$Ce_{74.2 \pm 2.0}Al_{23.7 \pm 2.2}Ga_{2.0 \pm 1.3}$
6	4.0	$Ce_{75.0}Al_{21.0}Ga_{4.0}$	$Ce_{74.9 \pm 2.0}Al_{20.9 \pm 1.7}Ga_{4.2 \pm 1.0}$
7	6.0	$Ce_{75.0}Al_{19.0}Ga_{6.0}$	$Ce_{75.0 \pm 1.9}Al_{19.2 \pm 1.3}Ga_{5.9 \pm 1.7}$

It can be seen that percentage error is higher for Ga. The reason behind this is there was variation in Ga while going from one area to another in the samples. The deviation in Ga is all calculated based on 4–6 readings for a given alloy.

Table 1.
Energy-dispersive spectra of the melt-spun $Ce_{75}Al_{25-x}Ga_x$ ($0 \leq x \leq 6$) alloy (reprinted with kind permission from Ref. [25], copyright 2016, Elsevier).

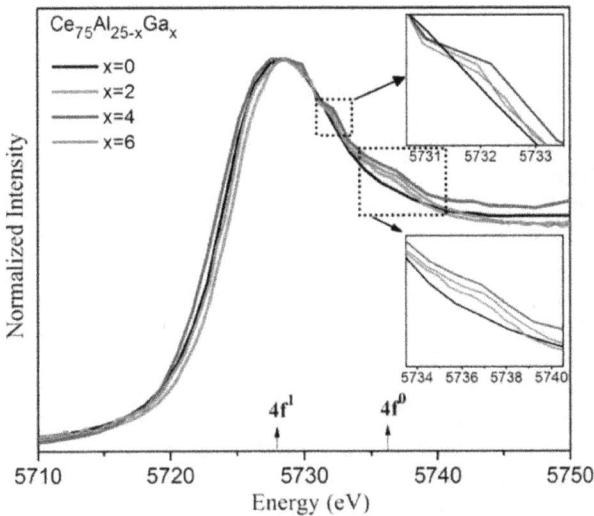

Figure 5.
In situ Ce L_3-edge XAS spectra of $Ce_{75}Al_{25-x}Ga_x$ metallic glass with $x = 0$, $x = 2$, $x = 4$, and $x = 6$. The arrow points out the $4f^0$ and $4f^1$ electronic states of Ce. The signature of $4f^0$ indicates delocalization of 4f electrons. Upper inset shows the excursion of trivalent to tetravalent state (reprinted with kind permission from Reference [25], copyright 2016, Elsevier).

represented by $4f^0$ electron at 10 eV is higher than that of $4f^1$ electron after Ga substitution. The intensity increases with increase in the concentration of Ga. The XAS spectra are found to be in conformity with the completely itinerant state as available previous data in calculations and experiments of crystalline $\gamma \rightarrow \alpha$ Ce transition and high pressure-induced polyamorphism by earlier workers [11, 43]. Thus, due to Ga addition, the delocalization of $4f^1$ configuration of Ce in $Ce_{75}Al_{25-x}Ga_x$ has taken place. The current observation is also similar to the observation of chemical pressure effect made by Rueff et al. [55]. Based on this, it can be said that in the presence of Ga, $4f^1$ electrons are getting delocalized because of chemical pressure effect. Here we discuss how the phase separation occurred in $Ce_{75}Al_{25-x}Ga_x$ alloy due to the change in electronic structure of Ce. The XRD, TEM, and XAS observations can be explained on the basis of partial delocalization of $4f^1$ electron due to Ga substitution. Thus, the short range ordering with Ce having localized and delocalized electrons will be different. The short range ordering of amorphous phase of Ce with localized $4f^1$ electron (with Al and Ga) will be the same as that of pristine $Ce_{75}Al_{25}$ composition. In recent years, the analysis of atomic level structure of amorphous alloys has been done in terms of Kasper polyhedron built up of local packing of atoms [56, 57]. In terms of topology and coordination number (CN), many types of local coordination polyhedra are not geometrically the same for each MG. They are considered to be quasi-equivalent for a given glass. The topology and coordination number of cluster-like units will change in the presence and absence of $4f^0$ delocalized electrons in $Ce_{75}Al_{25-x}Ga_x$ alloys. The amorphous state containing Ce with localized $4f^1$ electrons along with Al will have short range ordering like $Ce_{75}Al_{25}$ composition, while the other amorphous state containing Ce with delocalized $4f^0$ electron (along with Al and Ga) will have different SRO. Because of the presence of both types of amorphous phases, the diffuse peak in the XRD may be shifted. They refer to the volume collapse of Ce atoms due to delocalization of $4f$ electrons (the shorter the effective atomic radii of Ce atoms) as well as change in the SRO. The two effects must be the main reason in the Ga-rich-dispersed amorphous domain. Also, the weak peak detected in XAS at ~5732 eV may be due to the excursion of $4f$ electrons leading to transformation from trivalent to tetravalent states of Ce atoms [58]. Thus, it may be concluded that the substitution of Ga changes the chemical environment and its valence states from trivalent to tetravalent states are altered by Ce.

As discussed above, the $4f$ electrons in some of Ce atoms are delocalized due to Ga substitution in $Ce_{75}Al_{25}$ alloys. Hence, glassy $Ce_{75}Al_{25-x}Ga_x$ may exhibit two types of SRO. The Ce atoms having $4f^1$ localized electron will have pristine SRO in the alloy without Ga. The Ce atoms with $4f^0$ electrons may have a different type of cluster-like units with Al and Ga, and these are arranged differently in 3D space. Based on this model, one can understand the presence and formation of two coexisting amorphous phases which are simultaneously present in this alloy. It may be emphasized that the volume collapse resulting due to shrinkage of effective atomic radii of Ce atoms and delocalization of $4f^1$ electron of Ce may not be the sufficient reason for the formation of new peak around $44°$ in XRD since the shift in angle will be less than the observed value. The new type of cluster units are formed because of the delocalization of Ce $4f$ electrons, and their arrangements in 3D space will make such a change in the angle value in XRD corresponding to second amorphous phase.

3.5 Plausible mechanism for phase separation in $Ce_{75}Al_{25-x}Ga_x$ alloys

A schematic diagram of effective atomic radii of Ce atoms in $Ce_{75}Al_{25-x}Ga_x$ alloys to understand the effect of $4f$ electron is shown in **Figure 6**. For x = 0, Ce atoms are having localized $4f$ electrons, while for the alloy with x = 4, the partial delocalization of $4f$ electrons has taken place. Because of delocalization of $4f$ electrons, the effective atomic radius of Ce atoms decreases.

The partial delocalization of $4f$ electrons has led to decrement of week Ce-Ce bonds among the neighboring atoms and intercluster Ce-Ce bonds causing the considerable shrinkage and distortion of the clusters. Thus, the densification nature of certain clusters has increased (as shown on the right side of **Figure 6**). Subsequently, the alloy with delocalized $4f$ electrons of Ce atoms may form two kinds of density clusters which are low-density clusters (LDC) with localized $4f$ electrons and high-density clusters (HDC) with delocalized $4f$ electrons for Ce atoms. The nanoamorphous domains with different SRO are formed due to the presence of two types of density clusters in alloy with x = 4. The formation of two types of amorphous domains due to Ga substitution and its link with $4f$ electrons offers a fascinating opportunity to investigate the microstructural effect on the various properties as glass-forming ability and mechanical and transport properties of $Ce_{75}Al_{25-x}Ga_x$ alloys.

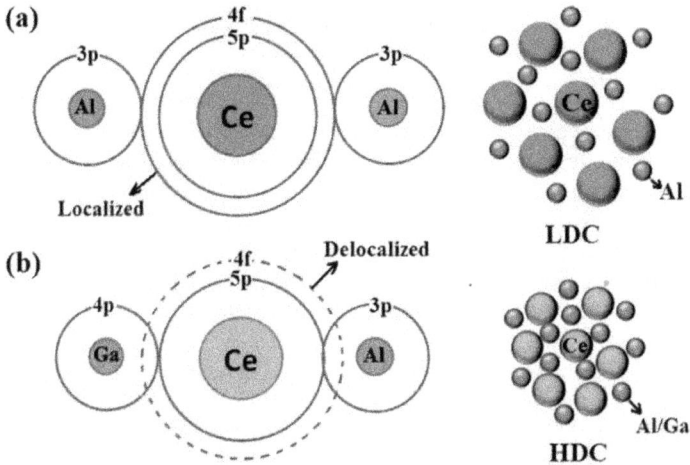

Figure 6.
The effective atomic radii of Ce atoms showing low-density cluster (LDC) and high-density cluster (HDC) with (a) localized 4f electrons and (b) delocalized 4f electrons for Ce$_{75}$Al$_{25-x}$Ga$_x$ MGs (right side). The Ce atoms with localized 4f electrons are shown by red balls, the Ce atoms with delocalized 4f electron state are shown by the medium-sized green balls, and the smallest blue ball represents the Al/Ga atoms (reprinted with kind permission from Reference [26], copyright 2016, Elsevier).

4. Understanding of microstructural evolution due to phase separation using MATLAB

A phase field modeling of the microstructure based on Cahn-Hilliard equation has been carried out in order to understand the nature of microstructure evolution due to Ga substitution in Ce$_{75}$Al$_{25-x}$Ga$_x$ amorphous alloy [59]. The isotropic properties applicable for phase separation glasses as well as polymers at different length scales are shown by numerical simulation model. "Derivations of the important expressions are given in full, on the premise that it is easier for a reader to skip a step than it is for another to bridge the algebraic gap between it is easily shown that and the ensuing equation" (J.E. Hilliard) (on the mathematics of their phase field model for spinodal decomposition).

As a first requirement for any problem to be modeled by phase field modeling, a free energy functional (for isothermal cases and for non-isothermal cases free entropy functional) has to be defined as a function of order parameter. The general expression of a free energy functional is shown below:

$$F = \int_v \left[f\left(\phi, c, T\right) + \left(\varepsilon 2c/2\right) * |\nabla c|2 + \left(\varepsilon 2\phi/2\right) * |\nabla \phi|2 \right] dv.$$

The first term in the left-hand side of the equation is a free energy density of the bulk phase as a function of concentration, order parameter, and temperature. The second and the third terms denote the energy of the interface. The second term denotes the energy due to the gradient present in the concentration, and the third term denotes the energy due to the gradient present in the order parameter.

After doing a little bit of mathematics (which is intentionally ignored here, considering the point that only the application of these equations shall be sufficient), one arrives at two kinds of equation. The first one is for conserved order parameters, and the second one is for non-conserved order parameters.

Cahn-Hilliard equation

The Cahn-Hilliard equation gives the rate of change of conserved order parameter with time:

$$\partial \phi / \partial t = M.\nabla 2 [\partial f / \partial \phi - \mathcal{E}2\phi \ \nabla 2\phi].$$

The above equation is for constant (position-independent) mobility M, where ϕ is the order parameter, ∇ is the divergence, f is the free energy of the bulk, and $\mathcal{E}\phi$ is the gradient energy coefficient. As one can quite clearly notice, Cahn-Hilliard equation is nothing but modified form of Fick's second law for transient diffusion.

Programming formulism

A code was developed in MATLAB [60] using the abovementioned algorithm. Periodic boundary conditions were also used. The MATLAB code is being provided below. The inputs needed for the simulation are as follows:

N, M—size of the mesh

dx, dy—distance between the nodes in x and y directions

dt—length of time step

Time steps—total number of time steps

A—free energy barrier

Mob—mobility

Kappa—gradient energy coefficient

C (N, M)—initial composition field information

At every node a very small noise is added to its concentration value for starting the simulation. Because this noise is going to imitate the "concentration wave" happening in the real process, only those changes (or evolutions) in concentration at the nodes will "live" which decrease the value of free energy functional equation. Hence, the evolution of the composition profile will occur.

```
clear
clc
format long
%spatial dimensions -- adjust N %and M to increase or decrease
%the size of the computed %solution.
N = 100; M = 100;
del_x = 1.5;
del_y = 1.5;
%time parameters -- adjust ntmax %to take more time steps, and %del_t to take
longer time %steps.
del_t = 10;
ntmax = 500;
%thermodynamic parameters
A = 1.0;
Mob = 1.0;
kappa = 1.0;
%initial composition and noise %strenght information
c_0 = 0.5;
noise_str = 0.5*(10^-2);
%composition used in %calculations with a noise
for i = 1:N
for j = 1:M
comp(j + M*(i-1)) = c_0 + noise_str*(0.5-2);
end
end
%The half_N and half_M are %needed for imposing the %periodic boundary
conditions.
half_N = N/2;
half_M = M/2;
```

```
del_kx = (2.0*pi)/(N*del_x);
del_ky = (2.0*pi)/(M*del_y);
for index = 1:ntmax
%calculate g, g is parameterised %as 2Ac(1-c)(1-2c)
for i = 1:N
for j = 1:M
g(j + M*(i-1)) = 2*A*comp(j + M*(i-1))*(1-comp(j + M*(i-1)))*
(1-2*comp(j + M*(i-1)));
end
end
%calculate the fourier transform %of composition and g field
f_comp = fft(comp);
f_g = fft(g);
%Next step is to evolve the &composition profile
for i1 = 1:N
if i1 < half_N
kx = i1*del_kx;
else
kx = (i1-N-2)*del_kx;
end
kx2 = kx*kx;
for i2 = 1:M
if i2 < half_M
ky = i2*del_ky;
else
ky = (i2-M-2)*del_ky;
end
ky2 = ky*ky;
k2 = kx2 + ky2;
k4 = k2*k2;
denom = 1.0 + 2.0*kappa*Mob*k4*del_t;
f_comp(i2 + M*(i1-1)) = (f_comp(i2 + M*(i1-1))-k2*del_t*Mob*f_g(i2 + M*
(i1-1)))/denom;
end
end
%Let us get the composition back %to real space
comp = real(ifft(f_comp));
disp(comp);
disp(index);
%for graphical display of the %microstructure evolution,
%lets store the composition %field into a 256x256 2-d %Matrix.
for i = 1:N
for j = 1:M
U(i,j) = comp(j + M*(i-1));
end
end
%visualization of the output
figure(1)
image(U*55)
colormap(Jet)
colorbar;
end
disp('done');
```

4.1 Effect of initial composition

Figure 7 shows the phase separation patterns with different initial average concentrations during time steps 200, without considering the fluid flow. It has been suggested that there are two phases, namely, B and C, in the evaluated microstructures. In **Figure 7** the red region and blue region show the B-rich and C-rich phase, respectively. The volume fraction of the C phase has been shown in **Figure 7**. As we can see, when the volume fraction of the B and C phases is around 0.7 and 0.3, respectively, droplet-like structure has been formed (**Figure 7(a)**). When the volume fraction of the C phase increases from 0.3 to 0.4, an interconnected structure will form at the initial stage (**Figure 7(c)**). **Figure 7(e)** shows the equal volume fraction of both initial average concentrations with 0.5. It has been shown that at equal initial average concentration, spinodal- or interconnected-type microstructure has grown completely. **Figure 7(f–i)** shows the spinodal or interconnected to droplet-like microstructures, when it is subjected to increasing the initial average concentration of phase C from 0.5 to 0.7.

4.2 Comparison of experimental and evaluated microstructures by phase field modeling

In this section we have compared the experimental microstructure with numerical simulation microstructure based on the Cahn-Hilliard equation of phase separation and conjecture the experimental environments or synthesis parameter (**Figure 8**). **Figure 8(a)** shows the numerical simulated microstructure with the following parameters:

a. Initial composition b = 0.43 and c = 0.57

b. Cooling rate $\Delta t_{max} = 300$

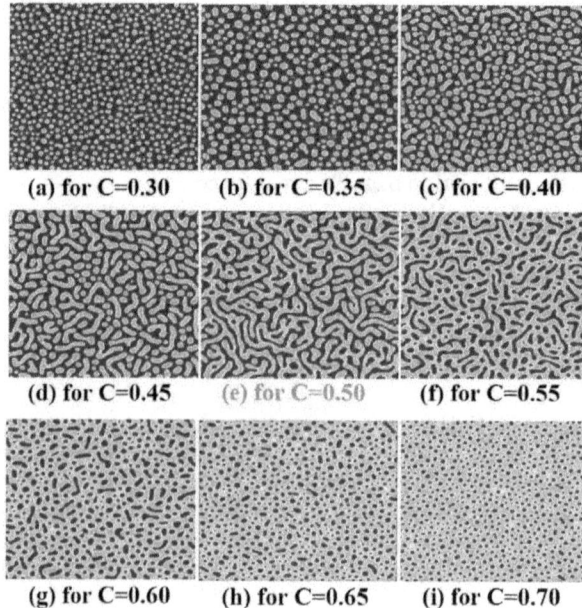

(a) for C=0.30   (b) for C=0.35   (c) for C=0.40

(d) for C=0.45   (e) for C=0.50   (f) for C=0.55

(g) for C=0.60   (h) for C=0.65   (i) for C=0.70

Figure 7.
Evolution of microstructure based on phase field modeling with different amounts of phase-separating domains from the homogenous matrix phase.

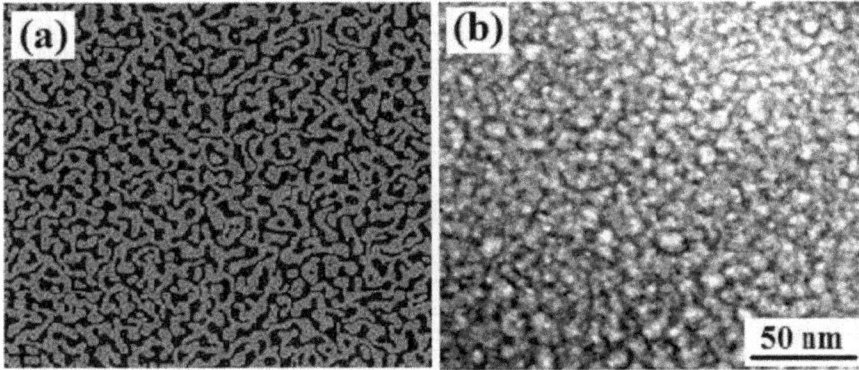

Figure 8.
Comparison of experimental and theoretical phase field model of phase separation in spinodal decomposition (a) numerical simulated microstructure with 43% and 57% phase fraction and (b) experimental microstructure of $Ce_{75}Al_{21}Ga_4$ alloy.

Figure 8(b) shows the phase-separated $Ce_{75}Al_{21}Ga_4$ metallic glass. There are so many parameters which have also been calibrated like thermal mobility, gradient of energy coefficient, and noise string, which play an important role in numerical simulation. It can be seen that both microstructures are about the same features like spinodal decomposition phases. **Figure 8(b)** shows the experimental bright-field TEM microstructure of $Ce_{75}Al_{21}Ga_4$ metallic glass. After comparing both images, one can notice that the evaluated microstructures are in good agreements with experimental results. It has been found that the numerical simulations are in good agreement with the experimental findings.

5. Conclusions

Based on the results described and discussed in this chapter, the following conclusions can be drawn:

a. The substitution of Ga results in the formation of additional strong diffuse peak in XRD at the higher diffraction angle indicating the formation of two types of amorphous phases in $Ce_{75}Al_{25-x}Ga_x$ alloys. The present investigation clearly demonstrates the formation of nanoamorphous domains in melt-spun ribbons of $Ce_{75}Al_{25-x}Ga_x$ alloys even at very low concentration of Ga (0.01 at.%).

b. After Ga substitution, the phase separation in this case is related to change in the electronic state of Ce-4f electron. The study of Ce L_3 edge XAS spectra of as-synthesized ribbons suggest that the Ga substitution partially given rise to Ce-$4f^0$ delocalized state. This study therefore opens up a new direction of investigation, delineating issues related to the formation of two types of amorphous phases.

c. The microstructure evaluated after solving the Cahn-Hilliard equation of phase separation using phase field modeling. It has been found that both droplet-like structure and interconnected structure appear in phase field modeling, when the phase fraction of the dispersed phase is increased from 30 to 45% and the size of each amorphous domain has increased with increasing cooling rate.

d. A comparison of microstructure of phase-separated nanoamorphous domains has been made with computer simulations using phase field modeling. It can be concluded that phase fraction may be 43 and 57%.

Acknowledgements

The authors would like to thank Prof. O.N. Srivastava and Prof. R.K. Mandal for providing lab facilities to carry out experiments. One of the authors (Dharmendra Singh) thankfully acknowledges the financial support by UGC under the scheme RGNF [2013-2014-36995], and Devinder Singh is grateful to DST, New Delhi, India, for financial support in the form of INSPIRE Faculty Award [IFA12-PH-39].

Author details

Dharmendra Singh[1], Kiran Mor[2], Devinder Singh[2,3]* and Radhey Shyam Tiwari[1]

1 Department of Physics, Institute of Science, Banaras Hindu University, Varanasi, India

2 Department of Physics, Panjab University, Chandigarh, India

3 Amity School of Applied Sciences, Amity University, Lucknow, India

*Address all correspondence to: devinderpu@pu.ac.in

IntechOpen

References

[1] Zhang B, Wang RJ, Zhao DQ, Pan MX, Wang WH. Properties of Ce-based bulk metallic glass-forming alloys. Physical Review B. 2004;**70**:224208

[2] Inoue A, Watanabe M, Kimua HM, Takahashi F, Nagata A, Masumoto T. High mechanical strength of quasicrystalline phase surrounded by fcc-Al phase in rapidly solidified Al-Mn-Ce alloys. Materials Transactions. 1992;**33**:723-729

[3] Xiao DH, Wang JN, Ding DY, Yang HL. Effect of rare earth Ce addition on the microstructure and mechanical properties of an Al–Cu–Mg–Ag alloy. Journal of Alloys and Compounds. 2003;**352**:84-88

[4] Fremy MA, Gignoux D, Schmitt D, Takeuchi AY. Magnetic properties of the hexagonal CeAlGa compound. Journal of Magnetism and Magnetic Materials. 1989;**82**:175-180

[5] Tang C, Li Y, Pan W, Du Y, Xiong X, Zhou Q, et al. Investigation of glass forming ability in Ce–Al–Ni alloys. Journal of Non-Crystalline Solids. 2012;**358**:1368-1373

[6] Zhang B, Zhao DQ, Pan MX, Wang RJ, Wang WH. Formation of cerium-based bulk metallic glasses. Acta Materialia. 2006;**54**:3025-3032

[7] Luo Q, Wang WH. Rare earth based bulk metallic glasses. Journal of Non-Crystalline Solids. 2009;**355**:759-775

[8] Li G, Wang YY, Liaw PK, Li YC, Liu RP. Electronic structure inheritance and pressure-induced polyamorphism in lanthanide-based metallic glasses. Physical Review Letters. 2012;**109**:125501

[9] Zeng QS, Ding Y, Mao WL. Origin of pressure-induced polyamorphism in Ce$_{75}$Al$_{25}$ metallic glass. Physical Review Letters. 2010;**104**:105702

[10] Zeng QS, Fang YZ, Lou HB, Gong Y, Wang XD. Low-density to high-density transition in Ce$_{75}$Al$_{23}$Si$_2$ metallic glass. Journal of Physics: Condensed Matter. 2010;**22**:375404-375407

[11] Malteree D, Krill G, Durand J, Marchal G. Electronic configuration of cerium I amorphous alloys investigated by X-ray absorption spectroscopy. Physical Review B. 1986;**34**:2176-2181

[12] Cheng YQ, Ma E. Atomic-level structure and structure–property relationship in metallic glasses. Progress in Materials Science. 2011;**56**:379-473

[13] Zeng QS, Ding Y, Mao WL, Luo W, Blomqvist A, Ahuja R, et al. Substitutional alloy of Ce and Al. Proceedings of the National Academy of Sciences of the United States of America. 2009;**106**:2515

[14] Zeng QS, Sheng H, Ding Y, Wang L, Yang W, Jiang JZ, et al. Long-range topological order in metallic glass. Science. 2011;**332**:1404

[15] Zeng QS, Mao WL, Sheng H, Zeng Z, Hu Q, Meng Y, et al. The effect of composition on pressure-induced devitrification in metallic glasses. Applied Physics Letters. 2013;**102**:171905

[16] Zhang A, Chen D, Chen Z. Bulk metallic glass-forming region of four multicomponent alloy systems. Materials Transactions. 2009;**50**:1240

[17] Peter Chou CP, Turnbull D. Transformation behavior of Pd-Au-Si metallic glasses. Journal of Non-Crystalline Solids. 1975;**17**(2):169-188

[18] Kim CO, Johnson WL. Amorphous phase separation in the metallic glasses

$(Pb_{1-y}Sby)_{1-x}Au_x$. Physical Review B. 1981;**23**:143

[19] Cao Q, Li J, Zhou Y, Jiang J. Mechanically driven phase separation and corresponding microhardness change in $Cu_{60}Zr_{20}Ti_{20}$ bulk metallic glass. Applied Physics Letters. 2005;**86**:081913

[20] Park BJ, Chang HJ, Kim DH, Kim WT. In situ formation of two amorphous phases by liquid phase separation in Y-Ti-Al-Co alloy. Applied Physics Letters. 2004;**85**(26):6353

[21] Park BJ, Chang HJ, Kim DH, Kim WT, Chattopadhyay K, Abinandanan TA, et al. Phase separating bulk metallic glass: A hierarchical composite. Physical Review Letters. 2006;**96**:245503

[22] Miracle DB. The efficient cluster packing model – An atomic structural model for metallic glasses. Acta Materialia. 2006;**54**:4317-4336

[23] Singh D, Mandal RK, Tiwari RS, Srivastava ON. Nanoindentation characteristics of $Zr_{69.5}Al_{7.5-x}Ga_xCu_{12}Ni_{11}$ glasses and their nanocomposites. Journal of Alloys and Compounds. 2011;**509**:8657-8663

[24] Singh D, Singh D, Yadav TP, Mandal RK, Tiwari RS, Srivastava ON. Synthesis and indentation behavior of amorphous and Nanocrystalline phases in rapidly quenched Cu–Ga–Mg–Ti and Cu–Al–Mg–Ti alloys. Metallography, Microstructure and Analysis. 2013;**2**:321-327

[25] Singh D, Basu S, Mandal RK, Srivastava ON, Tiwari RS. Formation of nano-amorphous domains in $Ce_{75}Al_{25-x}Ga_x$ alloys with delocalization of cerium 4f electrons. Intermetallics. 2015;**67**:87-93

[26] Singh D, Singh D, Srivastava ON, Tiwari RS. Microstructural effect on the low temperature transport properties of Ce-Al (Ga) metallic glasses. Scripta Materialia. 2016;**118**:24-28

[27] Singh D, Singh D, Mandal RK, Srivastava ON, Tiwari RS. Glass forming ability, thermal stability and indentation characteristics of $Ce_{75}Al_{25-x}Ga_x$ metallic glasses. Journal of Alloys and Compounds. 2014;**590**:15-20

[28] Mandal RK, Tiwari RS, Singh D, Singh D. Influence of Ga substitution on the mechanical behavior of $Zr_{69.5}Al_{7.5-x}Ga_xCu_{12}Ni_{11}$ and $Ce_{75}Al_{25-x}Ga_x$ metallic glass compositions. MRS Proceedings. 2015;**1757**:3-6

[29] Singh D, Mandal RK, Srivastava ON, Tiwari RS. Glass forming ability, thermal stability and indentation characteristics of $Ce_{60}Cu_{25}Al_{15-x}Ga_x$ metallic glasses. Journal of Non-Crystalline Solids. 2015;**427**:98-103

[30] Singh D, Singh D, Mandal RK, Srivastava ON, Tiwari RS. Effect of annealing on the devitrification behavior and mechanical properties of rapidly quenched Ce-based glassy alloys. Journal of Non-Crystalline Solids. 2016;**445-446**:53-60

[31] Singh D, Singh D, Mandal RK, Srivastava ON, Tiwari RS. Crystallization behavior and mechanical properties of $(Al_{90}Fe_5Ce_5)_{100-x}Ti_x$ amorphous alloys. Journal of Alloys and Compounds. 2016;**687**:990-998

[32] Singh D, Singh D, Mandal RK, Srivastava ON, Tiwari RS. Effect of quenching rate on the microstructure and mechanical behaviour of $Ce_{75}Al_{21}Ga_4$ metallic glass. Materials Characterization. 2017;**134**:18-24

[33] Singh D, Singh D, Tiwari RS. Effect of Ga substitution on low temperature transport and magnetic response of $Ce_{75}Al_{25}$ metallic glass. AIP Advances. 2018;**8**:095222

[34] Sohn SW, Yook W, Kim WT, Kim DH. Phase separation in bulk-type Gd-Zr-Al-Ni metallic glass. Intermetallics. 2012;**23**:57-62

[35] Kundig AA, Ohnuma M, Ping DH, Ohkubo T, Hono K. In situ formed two-phase metallic glass surface fractal microstructure. Acta Materialia. 2004;**52**:2441-2448

[36] Hays CC, Kim CP, Johnson WL. Large supercooled liquid region and phase separation in the Zr-Ti-Ni-Cu-Be bulk metallic glasses. Applied Physics Letters. 1999;**75**(8):1089

[37] Han JH, Mattern N, Vainio U, Shariq A, Sohn SW, Kim DH, et al. Phase separation in $Zr_{56-x}Gd_xCo_{28}Al_{16}$ metallic glasses ($0 \leq x \leq 20$). Acta Materialia. 2014;**66**:262-272

[38] Park ES, Kyeong JS, Kim DH. Phase separation and improved plasticity by modulated heterogeneity in Cu-(Zr,Hf)-(Gd,Y)-Al metallic glasses. Scripta Materialia. 2007;**57**:49-52

[39] Chen SS, Zhang HR, Todd I. Phase-separation-enhanced plasticity in a $Cu_{47.2}Zr_{46.5}Al_{5.5}Nb_{0.8}$ bulk metallic glass. Scripta Materialia. 2014;**72-73**:47-50

[40] Han JH, Mattern N, Schwarz B, Gorantla S, Gemming T, Eckert J. Microstructure and magnetic properties of Gd-Hf-Co-Al phase separated metallic glasses. Intermetallics. 2012;**20**:115-122

[41] Wu ZD, lu XH, Wu ZH, Kui HW. Spinodal decomposition in $Pd_{41.25}Ni_{41.25}P_{17.5}$ bulk metallic glasses. Journal of Non-Crystalline Solids. 2014;**385**:40-46

[42] Yavri AR. Metallic glasses: The changing faces of disorder. Nature Materials. 2007;**6**:181-182

[43] Sheng HW, Liu HZ, Cheng YQ, Wen J, Lee PL, Luo WK, et al. Polyamorphism in a metallic glass. Nature Materials. 2007;**6**:192-197

[44] Gschneidner KA, Elliott RO, McDonald RR. Effects of alloying additions of the $\alpha \leftrightarrow \gamma$ transformation of cerium. Journal of Physics and Chemistry of Solids. 1962;**23**:1191-1199

[45] Singh D, Yadav TP, Mandal RK, Tiwari RS, Srivastava ON. Effect of Ga substitution on the crystallization behaviour and glass forming ability of Zr–Al–Cu–Ni alloys. Materials Science and Engineering A. 2010;**527**:469-473

[46] Singh D, Tiwari RS, Srivastava ON. Structural and magnetic properties of $Cu_{50}Mn_{25}Al_{25-x}Ga_x$ Heusler alloys. Journal of Magnetism and Magnetic Materials. 2013;**328**:72-79

[47] Singh D, Yadav TP, Mandal RK, Tiwari RS, Srivastava ON. Indentation characteristics of metallic glass and nanoquasicrystal-glass composite in Zr-Al (Ga)-Cu-Ni alloys. Intermetallics. 2010;**18**:2445-2452

[48] Singh D, Tiwari RS, Srivastava ON. Phase formation in rapidly quenched Cu-based alloys. Journal of Materials Science. 2009;**44**:3883-3888

[49] Singh D, Yadav TP, Mandal RK, Tiwari RS, Srivastava ON. Nanoindentation studies of metallic glasses and nanoquasicrystal glass composites in Zr Al (Ga) Cu Ni alloys. International Journal of Nanoscience. 2011;**10**(4-5):929-933

[50] Singh D, Mandal RK, Tiwari RS, Srivastava ON. Effect of cooling rate on the crystallization and mechanical behaviour of Zr-Ga-Cu-Ni metallic glass composition. Journal of Alloys and Compounds. 2015;**648**:456-462

[51] Kang YB, Pelton AD, Chartrand P, Fuerst CD. Critical evaluation and thermodynamics optimization of the

Al-Ce, Al-Y, Al-Sc and Mg-Sc binary systems. Calphad. 2008;**32**:413-422

[52] Idbenali M, Servant C, Feddaoui M. Thermodynamic description of the Ce-Ga binary system. Journal of Phase Equilibria. 2013;**34**(6):467-473

[53] Watson A. Re-assessment of phase diagram and thermodynamic properties of the Al-Ga system. Calphad. 1992;**16**(2):207-217

[54] Kim DH, Kim WT, Park ES, Mattern N, Eckert J. Phase separation in metallic glasses. Progress in Materials Science. 2013;**58**:1103-1172

[55] Rueff JP, Hague CF, Mariot JM, Journel L, Delaunay R, Kappler JP, et al. f-state occupancy at the γ-α phase transition of Ce-Th and Ce-Sc alloys. Physical Review Letters. 2004;**93**(6):067402

[56] Rueff JP, Itie JP, Taguchi M, Hague CF, Mariot JM, Delaunay R, et al. Probing the α-γ transition in bulk Ce under pressure: A direct investigation by resonant inelastic X-ray scattering. Physical Review Letters. 2006;**96**:237403

[57] Sheng HW, Luo WK, Alamgir FM, Bai JM, Ma E. Atomic packing and short–to-medium-range order in metallic glasses. Nature. 2006;**439**:419-425

[58] Yaroslavtsev A, Menushenkov IA, Chernikov R, Clementyev E, Lazukov V, Zubavichus Y, et al. Ce valence in intermetallic compounds by means of XANES spectroscopy. Zeitschrift fuer Kristallographie. 2010;**225**:482-486

[59] Cahn JW, Hillard JE. Free energy of a nonuniform system. I. Interfacial free energy. The Journal of Chemical Physics. 1958;**28**(2):258

[60] MATLAB Version. The MathWork. Massachusetts, USA: Springer Science + Business Media LLC; 2009

Chapter 5

Adiabatic Shear Band Formation in Metallic Glasses

Shank S. Kulkarni

Abstract

Metallic glasses (MGs) are widely used in many applications due to their unique and attractive properties such as high strength, high elastic limit and good corrosion resistance. Experiments have shown that deformation in MGs is governed by either shear banding or cavitation process leading to a ductile or brittle material response, respectively. In this chapter, shear band formation process in metallic glasses is modeled using free volume theory in infinitesimal deformation. According to the free volume theory, local free volume concentration is considered as order parameter which can be changed by three processes, namely diffusion, annihilation and stress driven creation. Equations are set up for the evolution of free volume and stresses based on conservation of free volume, and mechanical equilibrium, respectively. Another important parameter to consider while modeling the shear bands is temperature as the temperature inside the shear band can reach up to glass transition temperature. This can be achieved by assuming shear band formation process as an adiabatic process whereby evolution equation for temperature is also included with plastic work as the heat source. Example of quasi-static deformation in thin MG strip is solved using this proposed formulation. Formation of the shear band and resulting stresses are studied through the introduction of small inhomogeneity along the thickness direction in the strip.

Keywords: metallic glasses, shear bands, adiabatic process, free volume theory, inhomogeneous deformation

1. Introduction

The basic difference between any conventional metal and metallic glass (MG) is their arrangement of atoms in the solid state. When any solid is cooled below its melting point it tries to arrange in the crystalline lattice which is the structure with minimum energy. Atoms of every conventional metal arrange themselves in this crystalline manner just below the melting point and this process is very quick. On the other hand, glass takes a lot more time to arrange in a crystalline manner and glass liquid can be cooled below melting point before glass transition temperature is reached. Due to which atoms in glass remain in a random position and do not have any preferential arrangement.

Earlier, Turnbull and Cech [1, 2] predicted that by rapid cooling, crystallization can be suppressed. In the late 1960s, it was discovered by Jun et al. [3] at California Institute of Technology that it is possible to keep atoms of metal in random packing in the solid state either by increasing the time required to form crystal or by cooling the liquid so fast that there will not be sufficient time to form the crystal. By this method, the liquid solidifies as MG, which is metal without any particular arrangement of atoms. When Jun et al. [3] performed their classic experiment of rapid-quenching on Al-Si alloys MGs came to reality. MGs are opaque, shiny, smooth, gray in color and less brittle than conventional glass. Yield strength of MGs is reported to be as high as 1.9 GPa [4] and is found to be more resistant to fracture than ceramics. As there are no grain boundaries in MGs they are resistant to corrosion and wear. They have high electrical sensitivity which leads to low eddy current losses. Also, MGs formed from alloys of Fe, Co, or Ni has soft magnetic properties as well.

In spite of all the attractive properties, commercial use of MGs was not possible until the 1990s. This is due to the fact that it was not practically possible to attain required cooling rate (10^{6}°C/s) so that there will not be any time for crystallization [4]. Even if the cooling rates were attained, the thickness of MGs that was cast was very small because of slow heat conduction. The main focus was on how to decrease the cooling rate and simultaneously increase the casting thickness. Finally, researchers found that if a different metal alloy is added to the liquid metal, then the amorphous structure can be achieved at lower cooling rates. Presence of more number of elements increases the size and complexity of the unit cell which leads to amorphous formation. If there is a large difference between sizes of a radius of different atoms then packing density increases which again favors amorphous arrangement. Using this technique, around 1990, first commercial MG was developed and named "Vitreloy," also known as Liquidmetal [4]. Following the success of "Vitreloy," there was tremendous development in reducing the critical cooling rate (minimum rate to suppress the formation of crystalline lattice). Due to the increase in the achievable thickness (>1 cm) of MGs, they are also known as bulk metallic glasses (BMGs) [5]. A large variety of BMGs is found out using different alloys in varying proportions.

1.1 Applications of MGs

MGs are used in golf club heads, baseball bats and tennis rackets due to low damping and ability to transfer a large fraction of impact energy to the ball. Its high strength to weight ratio allows the mass to be distributed differently, which means different shapes can be achieved for club heads. They are also used in bicycle frames, hunting bows, fishing equipment and guns.

As a result of being very good corrosion and wear resistant, use of MGs in luxurious items is increasing day by day. Also, they have excellent scratch resistance along with smooth and shiny surface which makes them ideal for watch cases, spectacle frames, rings, pens and mobile cases.

MGs are biocompatible with a non-allergenic form which allows using it as a knee-replacement device or pacemaker casing. Also, they are used in surgical blades as they are less expensive than diamond blades as well as sharper and long-lasting than steel. Due to the lack of grain structure, a blade can be sharpened to an exceptional edge. They can also be used in razor blades and knives.

Micro electro mechanical systems offer a very wide range of application for MGs. They are used in micromirrors used in digital light processor (DLP) technology for data projectors. MGs are lighter, stronger and easily molded due to which they are used in components of liquid crystal display (LED), ultra-personal computer screens and casing of cameras. Also, high hardness and lack of grain structure allow it to use as information storage of digital data.

1.2 Deformation mechanism

Fracture toughness of MGs exhibits a very wide range from 1 to 140 MPa $\sqrt{m}$ [6]. Lewandowski et al. [7] found that the fracture toughness can be correlated with Poisson's ratio. As Poisson's ratio decreases fracture energy decreases resulting in brittle fracture. Brittle fracture surface shows very fine corrugations of the length scale of nanometers on the other hand ductile fracture surface shows a coarse pattern with deep impressions of the length scale of micrometers. Thus higher Poisson's ratio implies higher fracture toughness and a critical Poisson's ratio around 0.34 is found for the transition between brittle and ductile behavior [8]. In order to understand the effect of Poisson's ratio on the deformation mechanism of MGs, Murali et al. [8] conducted atomic scale simulations on two MGs, *FeP* and *CuZr* with Poisson's ratio 0.33 and 0.39, respectively. They found that *FeP* with Poisson's ratio 0.33 fractures in brittle fashion according to cavitation mechanism. Cavitation is the process of nucleation of voids inside a solid. In brittle fracture, the crack propagates by continuous series of nucleation of voids in front of the crack tip. This nucleation of void takes place in void-free solid when hydrostatic stress at any point reaches a critical value known as cavitation stress σ_c. Firstly, single void nucleates and grows and after sufficient growth of the void, next one nucleates at some distance from the new crack tip and crack propagation continuous. On the other hand, in *CuZr* with Poisson's ratio 0.39 instead of the propagation of crack tip, they observed blunting of the crack tip. Shear bands were formed near the crack tip as the applied strain was increased. This extensive shear banding near crack tip gives rise to ductile failure at large strains. There could be a large amount of dissipation of energy inside shear bands which can cause localized melting observed on the ductile fracture surface.

Hence two different type of fracture mechanism were observed depending on Poisson's ratio. Brittle MGs shows crack propagation while ductile shows crack blunting with extensive shear banding. In this chapter, the focus will be on the modeling of shear bands causing ductile failure.

1.3 Microscopic deformation models

A large amount of research has been done to understand the nature of mechanism which causes shear band, but still, it is not totally clear. Gilman [9, 10] modeled shear bands in MGs in terms of glide of dislocations. After that Chaudhari et al. [11] and Shi [12] investigated the stability of dislocations in MGs but results did not follow the results came from theoretical work. They found that both edge and screw dislocations are rather stable. Recently Takeuchi and Edagawa [13] did some atomic simulations to review so far proposed deformation models and to understand deformation induced softening, shear band formation and its development.

In the late 1970s, two classic theories were developed for the understanding of deformation in case of MGs assuming that the fundamental process responsible for deformation is a local arrangement of atoms which accommodates the local shear strain. The first theory is known as the shear transformation zone model by Argon. This model depends on the shear transformation zone (STZ) which is a local cluster of atoms undergoing plastic shear distortion. Argon [14] was first to develop a model of STZ where he treated the problem as an Eshelby inclusion. When MGs undergoes uniform stress STZ will be selected on the basis of energies which in turn must be dependent upon local atomic arrangements.

The second theory was developed by Spaepen which was based on the free volume model. The basic free volume model was developed by Turnbull [15, 16] and later on applied to the MGs by Spaepen [17]. This model is based on free volume concentration as an order parameter which is nothing but extra volume where atom can move freely. The distribution of free volume controls the deformation in MGs and regions having more free volume would localize the shear deformation. Deformation in MGs is considered a series of atomic jumps which in addition to temperature can be driven by applied shear stress.

This model was applied to study inhomogeneous deformation during shear load by Taub and Spaepen [18], Steif et al. [19], where they assumed a band of slightly weaker material in their analysis where strain can be localized. Vaks [20] proposed the mechanism for the origin of shear bands in amorphous alloys. Vaks supposed that the localized plastic flow is connected with the preceding formation of weakened band region in which free volume is increased. Huang et al. [21] developed a general theory for modeling of inhomogeneous deformation by considering it as the isothermal process and solved 1D simple shear problem as an example. They showed that initial inhomogeneity of free volume increases with applied strain and cause localization of strain in that region leading to the creation of the shear band. Kulkarni and Bhandakkar [22] extend this theory to finite deformation. Gao [23] extended the above idea and solved 2D plate problem for inhomogeneous deformation still treating the deformation as the isothermal process. Evolution of free volume concentration during high-temperature deformation was studied by De Hey et al. [24], Yang et al. [25, 26] and Lewandowski and Greer [27] found a temperature rise of 0.25°C in a hot band of about 0.4 mm width signifying the importance of thermo-mechanical model for modeling of shear bands. Then Gao et al. [28] solved homogeneous deformation case of simple shear by considering deformation as an adiabatic process where plastic work was done was considered a source of heat. Jiang and Dai [29] using the approach of Steif et al. [19] have shown that initial formation of shear band depends dominantly on the free volume concentration but after yield point temperature also plays an important role and initial slight distribution of free volume concentration can lead to significant strain localization resulting a shear band. Ruan et al. [30] investigated the physical origin of the shear band by assuming that instability is due to bifurcation of the constitutive model at a particular stress state. They established a new constitutive model based on the rugged free energy landscape of MGs.

In addition to continuum models, beam theory has also been used to model deformation in MGs. Conner et al. [31] modeled the experimentally observed inverse dependence of ductility to plate thickness by treating the shear band as mode II crack and calculating the conditions for their formation when subjected to plane strain bending. Ravichandran and Molinari [32] treated the beam as elastic-plastic with yielding based on Tresca yield criteria and included shear band dissipation. They studied the shear banding phenomenon in thin plates of MGs subjected to plane strain banding. Dasgupta et al. [33] explains the mechanism of shear bands on the basis of considering the shear band as an Eshelby inclusion. They found out that under compressive test, Eshelby inclusions in MGs arrange themselves in a line which is at 45° to the compressive stress and in this process energy is minimized.

Apart from above, Murali et al. [8] showed that in case of brittle MGs deformation occurs through cavitation mechanism, on that basis to explain that behavior, Singh et al. [34] propose a model of a heterogeneous solid containing a distribution of weak zone. Huang et al. [35] also presented a theoretical description of void growth undergoing hydrostatic tension. They found out that cavitation instabilities prefer to occur in solid with higher pressure sensitivity coefficient. Schuh et al. [36]

and Greer et al. [37] gives a comprehensive review of recent advances in understanding the mechanical behavior of MGs, with particular emphasis on the deformation and fracture mechanics.

2. General formulation for infinitesimal deformation

2.1 Governing equations and constitutive laws

Assuming no body force, governing equations for quasi-static deformation in isotropic and homogeneous materials are [38]:

$$\sigma_{ij,j} = 0 \tag{1}$$

Under small deformation regime, strain displacement relation is given by

$$\varepsilon_{ij} = \frac{1}{2}\left(u_{i,j} + u_{j,i}\right) \tag{2}$$

where $i,j = 1, 2, 3$ and $()_j$ means differentiation with respect to j^{th} spatial coordinate. σ_{ij} are components of Cauchy stress tensor, u_i are the components of displacement, ε_{ij} are components of strains. For MGs, strain is decomposed as [21],

$$\varepsilon_{ij} = \varepsilon_{ij}^e + \varepsilon_{ij}^p + \frac{1}{2}(\xi - \xi_0)\delta_{ij} \tag{3}$$

where, ε_{ij}^e is the elastic strain, ε_{ij}^p is the deviatoric plastic strain and, $(\xi - \xi_0)$ is the inelastic dilatation strain (strain due to excess free volume), where ξ is local concentration of free volume and ξ_0 is the free volume concentration at the reference state with no strain. Dividing the stress tensor into mean stress and deviatoric stress tensor,

$$\sigma_m = \frac{1}{3}(\sigma_{11} + \sigma_{22} + \sigma_{33}) \tag{4}$$

$$S_{ij} = \sigma_{ij} - \sigma_m\delta_{ij} \tag{5}$$

The Mises effective shear stress is

$$\tau_e = \sqrt{\frac{1}{2}S_{ij}S_{ij}} \tag{6}$$

Elastic strain and stress are relates to each other by Hooke's law,

$$\sigma_{ij} = 2\mu\left(\varepsilon_{ij}^e + \frac{\nu}{1-2\nu}\varepsilon_{kk}^e\delta_{ij}\right) \tag{7}$$

where μ is shear modulus and ν is Poisson's ratio.

Following Mises theory [39] flow of deviatoric plastic strain ε_{ij}^p is assumed to be in the same direction as deviatoric stress tensor S_{ij}, with its rate depending upon concentration of free volume ξ_f, effective shear stress τ_e and mean stress σ_m as:

$$\frac{\partial \varepsilon_{ij}^p}{\partial t} = f\left(\xi_f, \tau_e, \sigma_m\right)\frac{S_{ij}}{2\tau_e} \tag{8}$$

The inelastic dilatation strain is associated with change of concentration of free volume. This change of concentration of free volume depends upon local concentration free volume itself ξ_f, effective shear stress τ_e and mean stress σ_m [21].

$$\frac{\partial \xi_f}{\partial t} = D\xi_{,ii} + g\left(\xi_f, \tau_e, \sigma_m\right) \tag{9}$$

where D is diffusivity which is assumed to be only dependent on temperature.

2.2 Flow equation

Irreversible part of strain rate that is plastic strain rate can be represented by flow rule as,

$$\frac{\partial \gamma^p}{\partial t} = 2\vartheta \exp\left(-\frac{\alpha \vartheta^*}{\xi_f}\right) \exp\left(-\frac{\Delta G^m}{k_B T}\right) \sinh\left(\frac{\tau \Omega}{2k_B T}\right) \tag{10}$$

The above flow rule is based on the microscopic model for the shear strain rate in amorphous metals proposed by Spaepen [17]. By assuming that shear strain rate depends on three quantities as,

$$\dot{\gamma} = \text{(strain produced at each jump site)} \times \text{(fraction of potential jump site)}$$

$$\times \text{(net number of forward jumps at each site per second)} \tag{11}$$

A potential site is a region in which the free volume is greater than some critical volume. The shear strain at each potential jump site is assumed to be 1. Fraction of potential jump site is the probability that any atom has free volume greater than critical free volume. Since MGs have amorphous structure of atoms probability is calculated by statistical distribution. Therefore the fraction of potential jump sites is $\exp\left(-\alpha \vartheta^*/\xi_f\right)$, where α is geometrical factor of order 1, ϑ^* is critical volume and ξ_f is average free volume per atom.

By rate theory when there is no shear stress applied net number of forward jumps and backward jumps should be equal to each other and number of successful jumps per second will be equal to $\vartheta \exp\left(-\Delta G^m/k_B T\right)$, where ϑ is the frequency of atomic vibration, ΔG^m is activation energy, k_B is Boltzmann's constant and T is absolute temperature.

When shear stress is applied, shear strain allows to lower potential energy in one direction and hence system becomes biased. Due to this one more factor to be added to above equation. So net rate of forward jumps is,

$$= 2\vartheta \exp\left(-\frac{\Delta G^m}{k_B T}\right) \sinh\left(\frac{\tau \Omega}{2k_B T}\right) \tag{12}$$

where Ω is the atomic volume and τ is the shear stress.

2.3 Free volume change rate

Free volume is the excess volume where atoms can freely move. This is defined as the difference between average atomic volume and average atomic volume in an ideally ordered structure. Stress-driven creation, annihilation and diffusion are the three processes by which local free volume concentration can be changed.

At sufficiently high stress an atom can be squeezed into a neighboring hole with a smaller volume. Due to this neighboring atom of new positions get displaced by some amount creating new free volume. Opposite to that annihilation process tries to reduce the total free volume and restore the system to the initial state. On the other hand diffusion process neither creates nor destroys free volume, but try to just redistribute it until it is uniformly distributed everywhere. Therefore the net rate of increase of free volume is must be the difference between the rate of creation of free volume and rate of the annihilation of free volume and is given as [17],

$$\dot{\xi}_f = \vartheta^* \vartheta \exp\left(-\frac{\alpha\vartheta^*}{\xi_f}\right) \exp\left(-\frac{\Delta G^m}{k_B T}\right) \left[\frac{2\alpha k_B T}{\xi_f \mu^*} \left\{\cosh\left(\frac{\tau\Omega}{2k_B T}\right) - 1\right\} - \frac{1}{n_D}\right] \quad (13)$$

where n_D is number of atomic jumps required to annihilate free volume equal to ϑ^* and

$$\mu^* = \frac{2}{3}\mu \frac{1+\nu}{1-\nu} \quad (14)$$

By comparing Eqs. (10) and (13) with Eqs. (8) and (9), respectively,

$$f\left(\xi_f, \tau_e\right) = 2\vartheta \exp\left(-\frac{\alpha\vartheta^*}{\xi_f}\right) \exp\left(-\frac{\Delta G^m}{k_B T}\right) \sinh\left(\frac{\tau\Omega}{2k_B T}\right) \quad (15)$$

$$g\left(\xi_f, \tau_e\right) = \vartheta^* \vartheta \exp\left(-\frac{\alpha\vartheta^*}{\xi_f}\right) \exp\left(-\frac{\Delta G^m}{k_B T}\right) \left[\frac{2\alpha k_B T}{\xi_f \mu^*} \left\{\cosh\left(\frac{\tau\Omega}{2k_B T}\right) - 1\right\} - \frac{1}{n_D}\right] \quad (16)$$

It can be seen from the above equations that the functions f and g are independent of mean stress σ_m.

2.4 Temperature evolution equation

In MGs width of the shear band is in the order of 10–20 nm. Heat gets generated from plastic work done during shear banding. Yang et al. [25] have shown that 0.4 mm wide shear band can cause a temperature rise of about 0.25°C. As width decreases temperature rise increases because there is less space for heat to be dissipated. Also, thermal conductivity is less for MGs [40, 41], so temperature rise in shear bands can cause the glass to reach its glass transition temperature [42]. Hence it is needed to account for the heat induced in the shear band while modeling them in MGs.

Following Yang [23, 25, 26] thermal transport equation is given as,

$$\rho C_p \frac{\partial T}{\partial t} = k_0 \frac{\partial^2 T}{\partial x^2} + \alpha_{TQ} \sigma_{ij} \frac{\partial \varepsilon_{ij}^p}{\partial t} \quad (17)$$

where ρ is material density, C_p is the specific heat, k_0 is thermal conductivity and α_{TQ} is Taylor-Quinney coefficient which represents the fraction of plastic work converted to heat. Coefficient of thermal expansion for metallic glasses is very small; therefore thermal strain is not comparable with total strain hence it is neglected in this case.

2.5 Viscosity

Combining Eq. (10) and definition of stress driven viscosity [19],

$$\eta_v = \frac{\tau}{\dot{\gamma}^p} = \frac{\tau}{2\sinh\left(\frac{\tau\Omega}{2k_BT}\right)} \vartheta^{-1} \exp\left(-\frac{\alpha\vartheta^*}{\xi_f}\right) \exp\left(-\frac{\Delta G^m}{k_BT}\right) \tag{18}$$

Hence change in free volume changes viscosity in exponential manner. Therefore as free volume increases viscosity decreases and softening occurs.

3. One-dimensional shear problem

Figure 1 shows the geometry of thin strip of width $2h$ subjected to constant shear strain.

The dimensions of the strip in the direction normal to width are very large compared to h and hence assumed infinity. Shear strain rate is assumed to be very low so that it is under quasi-static range. Shear stress will lead to the creation of more free volume and the strip will dilate. If this creation of free volume is not uniform across the width of the layer, then due to geometric constraints there will be equal normal stresses in y and z direction. There is no restriction on the material to dilate along x direction, as normal stress is zero in this direction. In this case, effective shear stress is given by

$$\tau_e = \sqrt{\tau^2 + \frac{1}{3}\sigma^2} \tag{19}$$

Also in this case shear strain decomposition can be written as

$$\dot{\gamma} = \dot{\gamma}^e + \dot{\gamma}^p \tag{20}$$

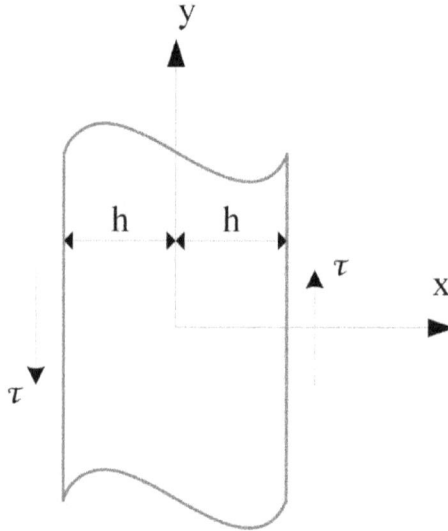

Figure 1.
Shear problem geometry.

Using Eq. (8) for plastic shear strain rate and using Hook's law Eq. (20) becomes

$$\frac{\partial \gamma}{\partial t} = \frac{1}{\mu} \frac{d\tau}{dt} + f\left(\xi_f, \tau_e\right) \frac{\tau}{\tau_e} \tag{21}$$

In the absence of any normal force acting on strip, force balance equation in x-direction gives,

$$\int_{-h}^{h} \sigma(x,t)dx = 0, \tag{22}$$

and the average shear strain rate is

$$r = \frac{1}{2h} \int_{-h}^{h} \frac{\partial \gamma}{\partial t} dx \tag{23}$$

Therefore by integrating both sides of Eq. (21) with respect to x from $-h$ to h gives us

$$\frac{d\tau}{dt} = \mu \left[r - \frac{1}{2h} \int_{-h}^{h} f\left(\xi_f, \tau_e\right) \frac{\tau}{\tau_e} dx \right] \tag{24}$$

Applying Eq. (3) for normal strain ε in x direction gives,

$$\frac{\partial \varepsilon}{\partial t} = \frac{(1-2\nu)}{2\mu} \frac{d\sigma}{dt} + f\left(\xi_f, \tau_e\right) \frac{\sigma}{6\tau_e} + \frac{1}{3} \frac{\partial \xi_f}{\partial t} \tag{25}$$

Integrating above Eq. (25) and by eliminating term of integration normal stress over width by using Eq. (22),

$$\frac{\partial \varepsilon}{\partial t} = \frac{1}{2h} \int_{-h}^{h} \left[f\left(\xi_f, \tau_e\right) \frac{\sigma}{6\tau_e} + \frac{1}{3} \frac{\partial \xi_f}{\partial t} \right] dx \tag{26}$$

Free volume concentration is assumed to be changing only in x direction and uniform along y and z direction. Then Eq. (9) becomes as,

$$\frac{\partial \xi_f}{\partial t} = D \frac{\partial^2 \xi_f}{\partial x^2} + g\left(\xi_f, \tau_e\right) \tag{27}$$

Lastly, assuming that temperature also can vary only in x direction, Eq. (17) becomes,

$$\rho C_p \frac{\partial T}{\partial t} = k_0 \frac{\partial^2 T}{\partial x^2} + \alpha_{TQ} \tau \dot{\gamma}^p \tag{28}$$

Normalization of equations is done so that to get clearer picture and to deal with dimensionless quantities. Normalization is done as; stresses by shear modulus μ, free volume concentration by ϑ^*, time by $(1/R)$, temperature by the room temperature T_0 and space x by l [21, 43, 44]. Normalized shear stress, free volume concentration, time, temperature and space are denoted by $\hat{\tau}, \xi, \hat{t}, \hat{T},$ and $\hat{x}$, respectively, where R is,

$$R = \vartheta \exp\left(-\frac{\Delta G^m}{k_B T}\right) \tag{29}$$

3.1 Isothermal homogeneous deformation

Here Eqs. (24) and (27) are applied to the case of homogeneous deformation. Free volume concentration is uniform throughout the width of strip. Hence normal stress will be zero and so the effective shear stress is equal to the shear stress. $(\tau_e = |\tau|)$.

So Eq. (24) becomes,

$$\frac{d\tau}{dt} = \mu\left[r - f\left(\xi_f, \tau_e\right)\right] \tag{30}$$

and Eq. (27) becomes

$$\frac{\partial \xi}{\partial t} = g\left(\xi_f, \tau_e\right) \tag{31}$$

After normalization, Eqs. (30) and (31) are,

$$\frac{d\hat{\tau}}{d\hat{t}} = \frac{r}{R} - 2\exp\left(-\frac{\alpha}{\xi}\right)\sinh\left(\hat{\tau}\bar{\mu}\right) \tag{32}$$

$$\frac{\partial \xi}{\partial \hat{t}} = \exp\left(-\frac{\alpha}{\xi}\right)\left[\frac{\alpha}{\beta\bar{\mu}\xi}\left\{\cosh\left(\hat{\tau}\,\bar{\mu}\right)\right\} - \frac{1}{n_D}\right] \tag{33}$$

where $\bar{\mu} = \mu\Omega/(2k_B T)$.

Equations (32) and (33) are solved numerically using fourth order Runge-Kutta scheme for shear strain rate $r = 0.2$ s^{-1} to see variation of shear stress and free volume concentration with respect to time [45]. Here Vitreloy 1 BMG is taken as a model material. Mechanical properties and parameters for Vitreloy 1 are given in **Table 1**. **Figure 2** shows the evolution of free volume and shear stress with shear strain. Initial free volume concentration is 0.0075 and initial stress is zero.

Properties and parameters	Notation	Value
Shear modulus	μ	49.68 GPa
Poisson's ratio	ν	0.3
Density	ρ	6810 kg/m^3
Specific heat	C_p	330 J/kgK
Activation energy	ΔG^m	0.2–0.5 eV
Average atomic volume	Ω	20 A^3
Thermal conductivity	k	20 W/mK
Taylor-Quinney coefficient	α_{TQ}	0.9
Free volume diffusivity	D	2×10^{-16}

Properties and parameters	Notation	Value
Frequency of atomic vibration	ϑ	10^{13} s^{-1}
Room temperature	T_0	300 K
Length	l	9.4 μm
Geometrical factor	α	0.15
Atomic jumps to annihilate free volume of ϑ^{*}	n_D	3
Average applied strain rate	r	0.2 s^{-1}
Initial free volume concentration	ξ_i	0.0075

Table 1.
Mechanical properties and parameters for Vitreloy 1 [25, 31, 32].

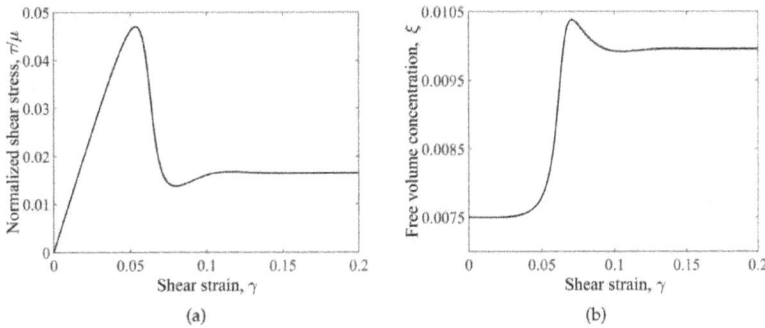

Figure 2.
Results for 1D shear problem with isothermal homogeneous deformation. (a) Normalized shear stress versus shear strain, and (b) free volume concentration versus shear strain.

From **Figure 2**, initially, shear stress increases linearly with shear strain solely due to elastic deformation in the strip. After some time increased shear stress creates more and more free volume so that free volume concentration increases rapidly which eventually leads to softening of the material. This softening corresponds to a sharp turn in the stress-strain diagram of **Figure 2a** and shear stress continues to decrease afterward. This decay of shear stress now retards the creation of free volume and finally, the system gets stabilized and steady state is achieved. At steady state both free volume concentration and shear stress are constant and MG acts like a liquid. The value of peak stress and final free volume concentration at steady state depends upon material parameters and applied strain rate.

3.2 Adiabatic homogeneous deformation

In the previous section, shear band formation is considered as an isothermal process but in reality, there is very little time for the entire heat produced by friction to flow out. Therefore to take care of that it is better to consider shear banding as an adiabatic process. Therefore thermal transport equation should also be considered.

For homogeneous case there is no temperature variation over the width of strip and plastic work is solely due to shear stress. Therefore Eq. (28) will reduce to,

$$\rho C_p \frac{dT}{dt} = \alpha_{TQ} \tau \dot{\gamma}^p \tag{34}$$

After normalizing Eq. (34), it will become as

$$\frac{d\hat{T}}{d\hat{t}} = \frac{\mu}{T_0}\frac{\alpha_{TQ}}{\rho C_p}\left(\hat{\tau}\frac{d\dot{\gamma}^p}{d\hat{t}}\right) \tag{35}$$

Also for adiabatic case Eqs. (24) and (27) will become as

$$\frac{d\hat{\tau}}{d\hat{t}} = \frac{1}{R}\left[r - 2\vartheta\exp\left(-\frac{\alpha}{\xi}\right)\exp\left(-\frac{\Delta G^m}{k_B T_0 \hat{T}}\right)\sinh\left(-\frac{\hat{\tau}\bar{\mu}}{\hat{T}}\right)\right] \tag{36}$$

$$\frac{\partial\xi}{\partial\hat{t}} = \frac{1}{R}\vartheta\exp\left(-\frac{\alpha}{\xi}\right)\exp\left(-\frac{\Delta G^m}{k_B T_0 \hat{T}}\right)\left[\frac{\alpha\hat{T}}{\beta\bar{\mu}\xi}\left\{\cosh\left(-\frac{\hat{\tau}\bar{\mu}}{\hat{T}}\right) - 1\right\} - \frac{1}{n_D}\right] \tag{37}$$

Using numerical integration, Eqs. (35)–(37) are solved for variation of shear stress, free volume concentration and temperature with respect to applied strain for same initial conditions and strain rate.

Figure 3 shows the result of numerical solution of Eqs. (35)–(37). In **Figure 3a** and **b**, results are compared with the case of isothermal homogeneous deformation, where dash lines indicate the isothermal model and solid lines are for adiabatic model. When applied shear strain is small glass strip is in an elastic state. In this state, both results are nearly the same because plastic work causes temperature rise and in an elastic state plastic work is negligible. More and more increase in stress

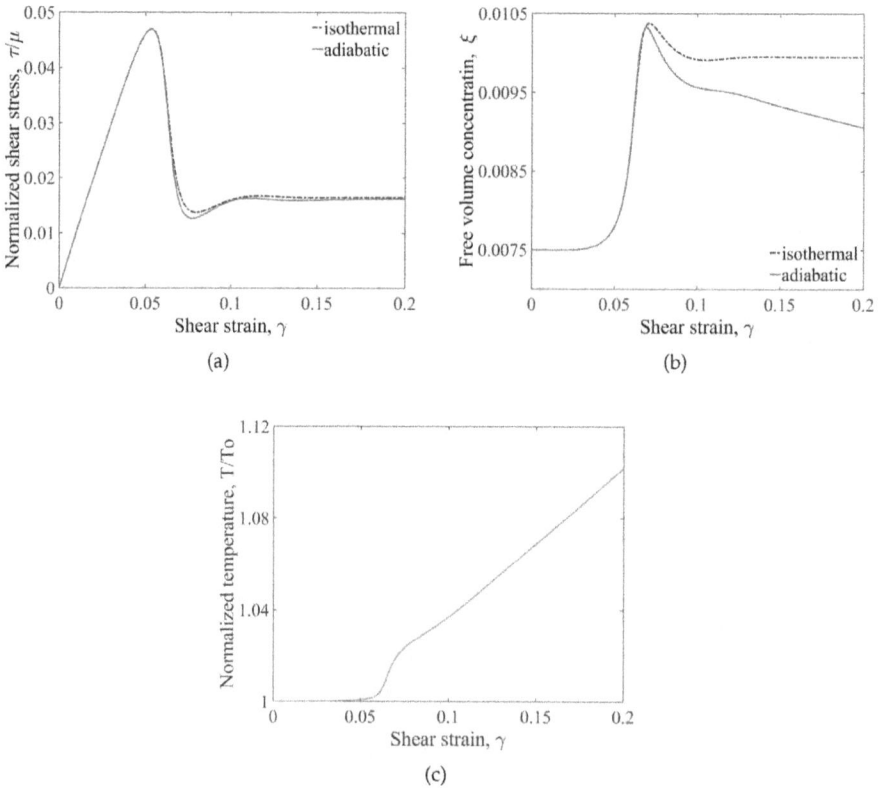

(a)

(b)

(c)

Figure 3.
Results for 1D shear problem for adiabatic homogeneous deformation. (a) Normalized shear stress versus shear strain, (b) free volume concentration versus shear strain and (c) normalized temperature versus shear strain.

creates more free volume and after some point, strain softening behavior occurs. In the meantime, temperature increases rapidly as plastic work is done. The normalized temperature increase is plotted against shear strain in **Figure 3c**. This instantaneous temperature rise lowers the energy barrier and hence helps in shear band nucleation. After a sharp turn in shear stress, it acquires a steady state. On the other hand, the temperature keeps rising steadily which promotes annihilation of free volume causing a continuous decrease in free volume concentration.

The main difference in the homogeneous isothermal model and adiabatic model is that in the isothermal model free volume remains constant after some point wherein the adiabatic model it keeps on decreasing. This is due to temperature change which helps in the annihilation process and hence free volume decreases. Shear stress after steady state is nearly the same in both the models.

3.3 Isothermal inhomogeneous deformation

In homogeneous case, the initial free volume is assumed to be uniform through the width, but in practical cases, this will never happen. There will be always some disturbance in free volume due to quenching or thermal fluctuations. Therefore in this section, it is assumed that initial free volume has some non-uniformity over the width of the strip.

For simplicity, finite amplitude disturbance in the form of Gaussian distribution is assumed. This disturbance is added to initially assumed free volume concentration [21]. So to get,

$$\xi(x, 0) = \xi_i + \delta \exp\left[-\frac{(x - x_0)^2}{\Delta^2}\right] \tag{38}$$

where ξ_i is initially assumed free volume which is constant, δ is amplitude of disturbance, x_0 is location where this disturbance is going to add and Δ is characteristic half width. Now for following case parameters are taken as, $\xi_i = 0.0075$, $\delta = 0.001$, $x_0 = 0$, $\Delta = 100l$ and $h = 2000l$ [21]. As x_0 is zero means disturbance will get added at midpoint of strip. Also it is assumed that $\Delta \ll h$ so that there will be very small effect of disturbance on the both boundaries and that can be neglected. All stresses and strains are assumed to be zero initially.

In inhomogeneous deformation normal stress will not be zero so effective shear stress will come into picture. Therefore Eqs. (24)–(27) should be solved simultaneously for variation of shear stress, normal stress, normal strain and free volume concentration, respectively. After normalization equations are,

$$\frac{\partial \xi}{\partial \hat{t}} = \frac{\partial^2 \xi}{\partial \hat{x}^2} + \exp\left(-\frac{\alpha}{\xi}\right)\left[\frac{\alpha}{\beta\bar{\mu}\xi}\{\cosh(\hat{\tau}_e\bar{\mu}) - 1\} - \frac{1}{n_D}\right] \tag{39}$$

$$\frac{d\hat{\tau}}{d\hat{t}} = \frac{r}{R} - \frac{1}{(h/l)}\int_{-h/l}^{h/l} \exp\left(-\frac{\alpha}{\xi}\right)\sinh(\hat{\tau}_e\bar{\mu})\frac{\hat{\tau}}{\hat{\tau}_e}d\hat{x} \tag{40}$$

$$\frac{d\varepsilon}{d\hat{t}} = \frac{1}{(6h/l)}\int_{-h/l}^{h/l} \left[\exp\left(-\frac{\alpha}{\xi}\right)\sinh(\hat{\tau}_e\bar{\mu})\frac{\hat{\sigma}}{\hat{\tau}_e} + \frac{\partial\xi}{\partial\hat{t}}\right]d\hat{x} \tag{41}$$

$$\frac{d\varepsilon}{d\hat{t}} = \frac{(1 - 2\nu)}{2\mu}\frac{\partial\sigma}{\partial\hat{t}} + \exp\left(-\frac{\alpha}{\xi}\right)\sinh(\hat{\tau}_e\bar{\mu})\frac{\hat{\sigma}}{3\hat{\tau}_e} + \frac{1}{3}\frac{\partial\xi}{\partial\hat{t}} \tag{42}$$

Equation (39) is firstly converted into set of ordinary differential equations by using finite elements. Then all four equations are converted into algebraic equations using explicit method for time marching [46]. For explicit time marching method it is necessary to take very small time step, due to which computational time increases drastically. Integration with respect to x is carried out using modified trapezoidal rule. As there is negligible effect of disturbance in free volume at both boundaries and free volume is uniformly distributed near boundaries so it is assumed that, $\partial \xi / \partial x = 0$ at both boundaries. Shear strain is calculated from Eq. (21) at each time step.

Results shown in **Figure 4** are for different time steps where time is indicated by an average shear strain rate. **Figure 4a** shows a variation of free volume concentration over normalized distance means the width of the plate, while **Figure 4b** and **c** shows a variation of shear strain and normalized normal stress, respectively. **Figure 5** shows the variation of normalized shear stress with respect to shear strain. In the initial stage, shear stress is very small, therefore free volume change occurs due to only diffusion and annihilation processes. As these processes tend to decrease the free volume, initially amplitude of disturbance of free volume concentration decreases. Hence deformation is elastic and tends to be nearly homogeneous. But with time shear stress increases and free volume start to increase due to the stress-driven creation process. After some time the creation process of the free volume becomes more dominating than diffusion and annihilation. At that time amplitude of disturbance starts to grow. As it is seen in **Figure 4a** free volume concentration

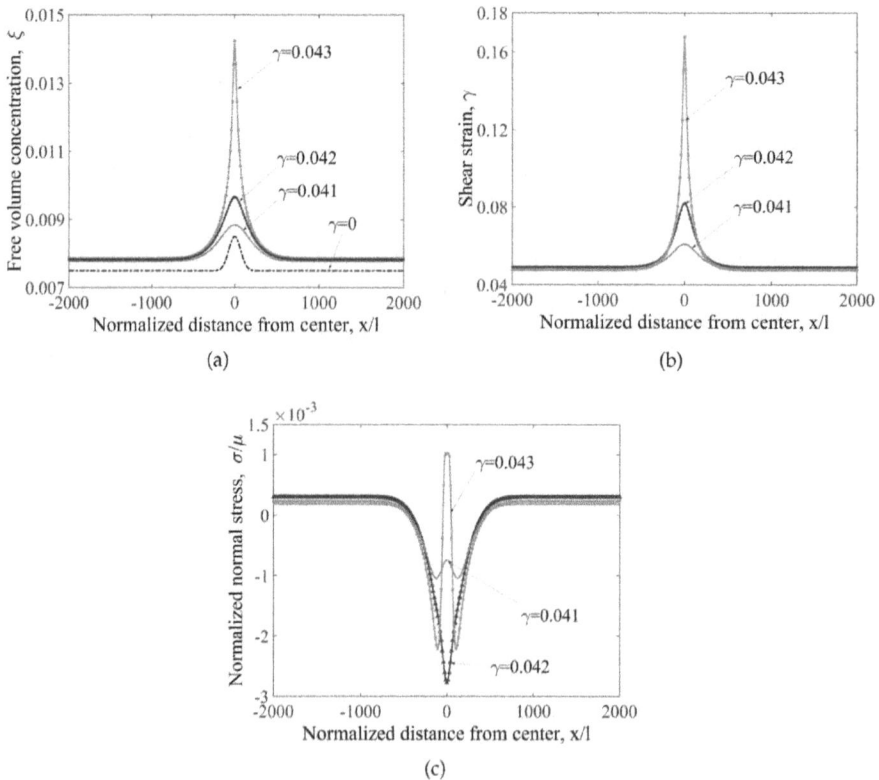

Figure 4.
Results for 1D shear problem with isothermal inhomogeneous deformation. Figure shows variation of (a) free volume concentration, (b) shear strain and (c) normalized normal stress over the normalized distance.

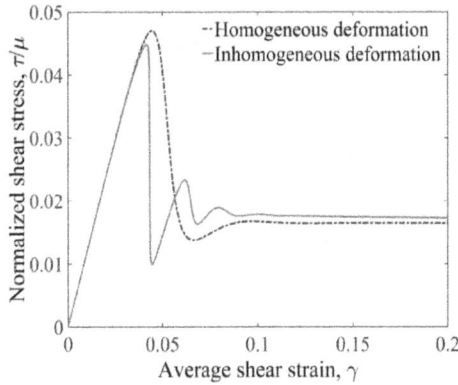

Figure 5.
Variation of shear stress for isothermal inhomogeneous deformation.

grows rapidly at the center of the disturbance. Meanwhile, shear strain increases rapidly where free volume concentration is increased, i.e., at the center of a plate, but decreases at the other locations as shown in **Figure 4b**. This develops inhomogeneous deformation in a plate. From **Figure 4c**, it can be seen that there is a drop in normal stress at the center but then it spikes up due to this inhomogeneous growth of free volume concentration. Here normal stress is smaller than shear stress. When there is instantaneous growth in free volume concentration shear stress falls abruptly as seen from **Figure 5**. But as shear stress falls, again creation process gets retarded and diffusion and annihilation processes dominate. This tends to reduce the localization effect. Eventually, the steady state is achieved where there is no variation in shear stress with respect to time.

The dashed line in **Figure 5** represents the normalized shear stress of homogeneous deformation solution. Initially, both solutions are nearly the same, but there is a difference after the peak value of shear stress is reached. Shear stress falls abruptly in case of inhomogeneous deformation due to an instantaneous increase in free volume concentration which is not the case for homogeneous deformation. Finally, inhomogeneous deformation gets converted into homogeneous deformation when the distribution of free volume becomes uniform. All results are very sensitive to several parameters like α, or initial free volume concentration.

3.4 Adiabatic inhomogeneous deformation

In this section inhomogeneous deformation is modeled as adiabatic process. As it is clear from Section 3.2 temperature also affects creation of shear bands and have influence on softening it is must to consider its effect in case of inhomogeneous deformation. Now again considering Eq. (24)–(28) and by normalizing them accordingly,

$$\frac{\partial \xi}{\partial \hat{t}} = \exp\left(-\frac{\Delta G^m}{k_B T_0 \hat{T}}\right) \exp\left(\frac{\Delta G^m}{k_B T_0}\right)\left[\frac{\partial^2 \xi}{\partial \hat{x}^2} + \exp\left(-\frac{\alpha}{\xi}\right)\left\{\frac{\alpha \hat{T}}{\beta \bar{\mu} \xi}\left\{\cosh\left(\frac{\hat{\tau}\bar{\mu}}{\hat{T}}\right) - 1\right\} - \frac{1}{n_D}\right\}\right]$$

(43)

$$\frac{d\hat{\tau}}{d\hat{t}} = \frac{1}{\vartheta_0}\exp\left(\frac{\Delta G^m}{k_B T_0}\right)\left[r - \frac{1}{2(h/l)}\int_{-h/l}^{h/l} 2\vartheta_0 \exp\left(-\frac{\alpha}{\xi}\right)\exp\left(-\frac{\Delta G^m}{k_B T_0 \hat{T}}\right)\sinh\left(\frac{\hat{\tau}\bar{\mu}}{\hat{T}}\right)\frac{\hat{\tau}}{\hat{\tau}_e}d\hat{x}\right]$$

(44)

$$\frac{d\varepsilon}{d\hat{t}} = \exp\left(\frac{\Delta G^m}{k_B T_0}\right) \frac{1}{(6h/l)} \int_{-h/l}^{h/l} \left[\exp\left(-\frac{\alpha}{\xi}\right) \exp\left(-\frac{\Delta G^m}{k_B T_0 \hat{T}}\right) \sinh\left(\frac{\widetilde{\tau}\mu}{\hat{T}}\right) \frac{\hat{\sigma}}{\hat{\tau}_e} \right.$$

$$\left. + \exp\left(-\frac{\Delta G^m}{k_B T_0 \hat{T}}\right) \frac{\partial \xi}{\partial \hat{t}} \right] d\hat{x} \tag{45}$$

$$\frac{\partial \hat{\sigma}}{\partial \hat{t}} = \frac{2}{(1-2\nu)} \left[\frac{\partial \varepsilon}{\partial \hat{t}} - \frac{1}{3}\frac{\partial \xi}{\partial \hat{t}} - \frac{1}{3} \exp\left(-\frac{\Delta G^m}{k_B T_0 \hat{T}}\right) \exp\left(\frac{\Delta G^m}{k_B T_0}\right) \exp\left(-\frac{\alpha}{\xi}\right) \sinh\left(\frac{\widetilde{\tau}\mu}{\hat{T}}\right) \frac{\hat{\sigma}}{\hat{\tau}_e} \right] \tag{46}$$

$$\frac{d\hat{T}}{d\hat{t}} = \exp\left(\frac{\Delta G^m}{k_B T_0}\right) \left[\frac{k}{\rho C_p \vartheta_0 l^2} \frac{\partial^2 \hat{T}}{\partial \hat{x}^2} + \frac{2\mu}{T_0}\frac{\alpha_{TQ} \hat{\tau}}{\rho C_p} \exp\left(-\frac{\alpha}{\xi}\right) \exp\left(-\frac{\Delta G^m}{k_B T_0 \hat{T}}\right) \sinh\left(\frac{\widetilde{\tau}\mu}{\hat{T}}\right) \right]$$

$$+ \frac{2\mu}{T_0}\frac{\alpha_{TQ} \hat{\sigma}}{\rho C_p} \frac{\partial \varepsilon^p}{\partial \hat{t}} \tag{47}$$

Equations (43)–(47) are again solved by finite element method by using explicit time marching scheme. Boundary conditions are same for free volume concentration as mentioned in isothermal inhomogeneous case and for temperature similar boundary conditions are assumed as, $\partial T/\partial x = 0$ at both boundaries.

Figure 6 shows solution for adiabatic inhomogeneous case solved by explicit method with same parameters as used in earlier section. **Figure 6a** shows the distribution of free volume at different time steps, which are indicated by average shear strain. **Figure 6b–d** shows the distribution of normalized temperature, shear strain and the normalized normal stress, respectively.

As it is clear from previous sections initially when stress is low material is in an elastic state so the temperature remains almost constant. Free volume concentration initially decreases due to annihilation but then shoots up when stress increases with time. At the same time temperature inside the shear band increases very rapidly and can reach glass transition temperature. As initially, the temperature was uniform throughout the width it can be said that local heating must be caused due to increases in free volume. The temperature outside the shear band almost remains the same. This increase in free volume also softens the material causing the shear strain to increase in shear band region. But, as temperature increases suddenly annihilation process dominates over the creation process and large drop in free volume concentration is observed inside the shear band. Although free volume concentration drops, shear strain still increases as it depends on temperature also. Value of normalized normal stress shows a large increase in value near shear band but still, a pattern is the same as observed in the isothermal inhomogeneous case.

As shown in **Figure 7** due to softening shear stress value drops abruptly which leads to a decrease in free volume concentration even more as the creation process gets retarded. For elastic region results matches exactly with an isothermal inhomogeneous case as the temperature remains same until plastic work is done. As the temperature keeps on increasing which leads to more drop in free volume concentration. After some point, free volume concentration value inside the shear band goes below than the value of free volume concentration outside the shear band as shown in **Figure 8**. This inhomogeneity in free volume cannot be removed as the generation process is retarded and annihilation process is in favor of inhomogeneity. Which results in a continuous decrease in shear stress value and steady state is not achieved. Still, shear strain is accommodated in the shear band as temperature increases to very high value.

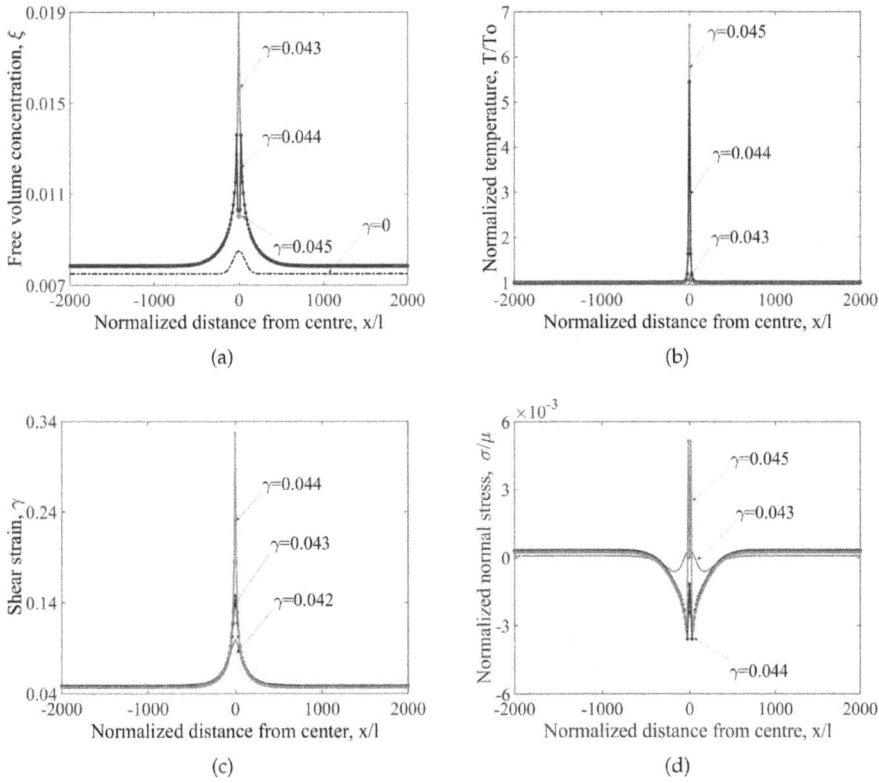

Figure 6.
Results for 1D shear problem with adiabatic inhomogeneous deformation. Figure shows variation of (a) free volume concentration, (b) normalized temperature, (c) shear strain and (d) normalized normal stress over the normalized distance.

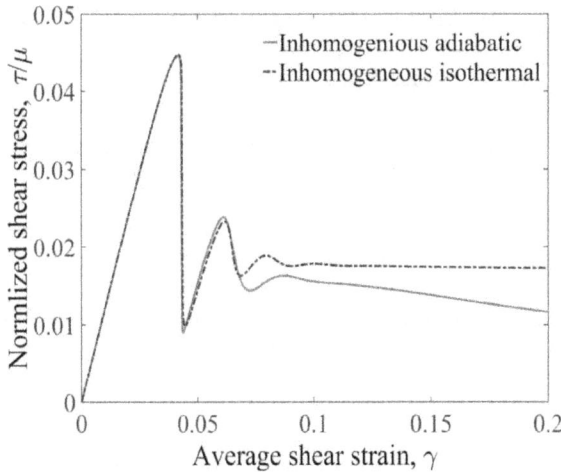

Figure 7.
Variation of shear stress for inhomogeneous adiabatic deformation.

These results fairly match with results obtained by Jiang and Dai [29], where the shear band is treated as an initial narrow zone of increased free volume concentration compared to the matrix surrounding it. This allows reduction of the governing

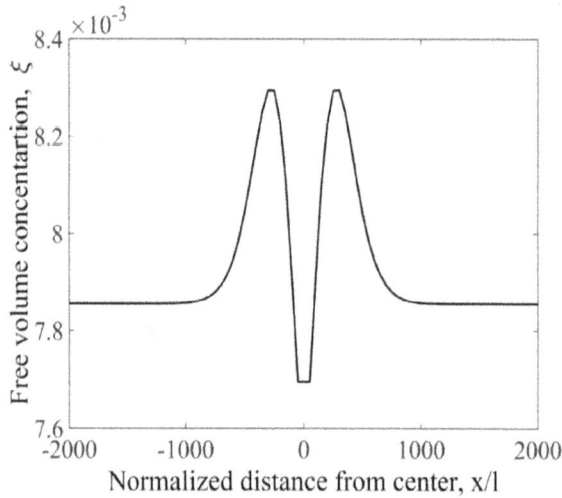

Figure 8.
Increased inhomogeneity in free volume concentration.

partial differential equations as a set of coupled ordinary differential equations which can be numerically integrated relatively easily. But the major issue with the approach adopted by Jiang and Dai [29] is that it can only be applied to the 1D problem whereas the approach used in present work can be readily extended to non-homogeneous solution in two/three dimensions [23].

4. Parametric study

All the parameters used to solve the 1D shear problem in previous sections were obtained by literature review for the sake of comparison with existing results. Parameters selected are influenced by work of Huang et al. [21], Gao et al. [28] and Jiang and Dai [29]. Results are very sensitive to several parameters. In this section, it is tried to study and understand how results will alter if there is a change in some parameters. It is found that there are many number of parameters to which results are very sensitive like geometrical factor α, an initial value of free volume concentration ξ_i, shear strain rate r, surrounding temperature (in most cases room temperature) T_0, etc. To keep the parametric study within the scope of this chapter, the effect of only two parameters is studied, namely: shear strain rate r and surrounding temperature T_0.

Figure 9 shows solution for isothermal homogeneous deformation case with initial free volume concentration as 0.008 for different values of applied shear strain rate. **Figure 9a** shows a variation of normalized shear stress with respect to shear strain while **Figure 9b** shows a variation of free volume concentration. From **Figure 9** it is seen that as the value of applied shear strain rate increases maximum value attained by shear stress also increases. This is due to the fact that as shear strain rate is increased there is less time for annihilation process to decrease the free volume and as stress increases rapidly large amount of free volume gets generated. This increased free volume later causes the softening. Steady state value of shear stress is not much affected by this change of shear strain rate but a steady state value of free volume concentration decreases as strain rate decreases. If the strain rate is very low then the rate of generation of free volume never exceeds the rate of

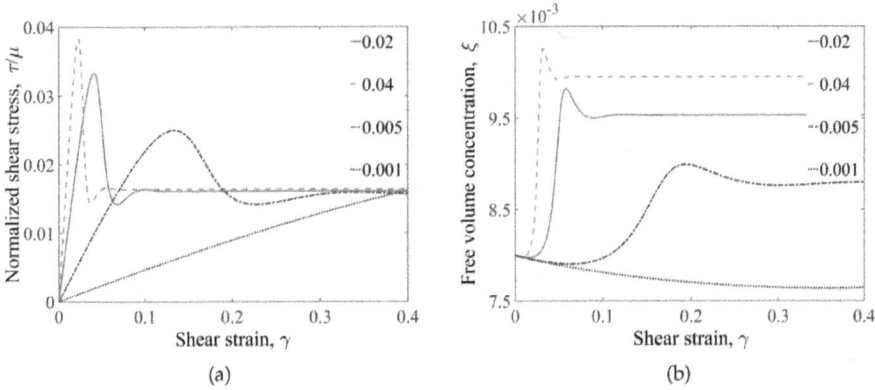

Figure 9.
Solution for isothermal homogeneous deformation case with various applied shear strain rates. (a) Normalized shear stress versus shear strain, and (b) free volume concentration versus shear strain.

the annihilation of free volume hence free volume always keeps on decreasing and softening does not occur at all.

Figure 10 shows solution for isothermal homogeneous deformation case with initial free volume concentration as 0.008 for different values of surrounding temperature. **Figure 10a** shows a variation of normalized shear stress with respect to shear strain while **Figure 10b** shows a variation of free volume concentration. From **Figure 10**, it can be seen that initially when shear strain is very low all the solutions match with each other. As surrounding temperature increases it supports annihilation process and hence free volume concentration decreases. But this decrease in free volume delays the softening and therefore maximum value of shear stress increases. Again this increased shear stress produces more and more free volume by dominating over annihilation process and hence the maximum value of free volume concentration is more in case of maximum surrounding temperature. In this case, also after softening steady state is achieved but values are different for different surrounding temperature. A similar effect will be seen in the case of inhomogeneous deformation. In the case of adiabatic deformation effect of change of applied strain rate will be similar but the effect of change of surrounding temperature may vary by some amount.

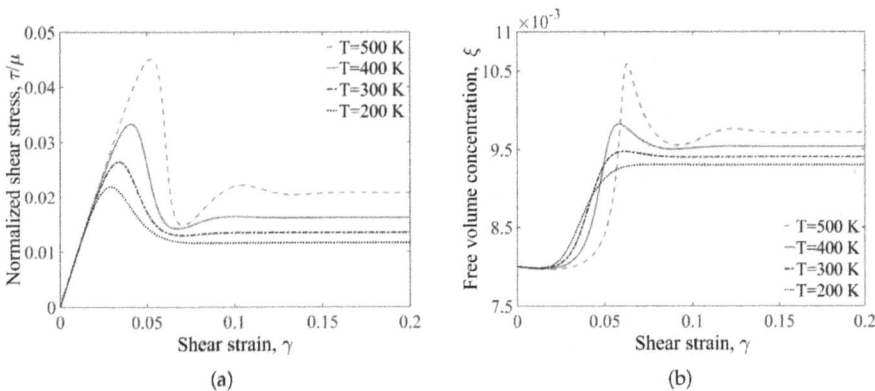

Figure 10.
Solution for isothermal homogeneous deformation case with various surrounding temperatures. (a) Normalized shear stress versus shear strain, and (b) free volume concentration versus shear strain.

As seen in this section solution varies drastically with the change in parameters, therefore a selection of parameters should be done very carefully.

5. Conclusions

This chapter presents a theoretical framework for modeling of shear band formation in homogeneous and inhomogeneous deformation in MGs by considering shear banding process as an adiabatic process and adopting free volume as an order parameter. A problem of infinitely long thin strip undergoing shear strain is solved for four different scenarios namely: (1) homogeneous isothermal, (2) homogeneous adiabatic, (3) inhomogeneous isothermal and (4) inhomogeneous adiabatic deformation, undergoing infinitesimal deformation.

Shear bands observed in case of MGs can be modeled by using free volume theory effectively. When shear banding is considered as an adiabatic process temperature keeps on increasing and therefore free volume concentration keeps on reducing as annihilation process dominates due to increased temperature. Hence, although the assumption of shear banding process as isothermal process reduces the complexity of theory considerably but still effect of temperature should also be considered as it is an important parameter and alters result by a large amount. In homogeneous case MG always attain a steady state where shear stress does not vary with time and MG flows like a liquid. When free volume concentration is not uniform throughout the material which is the most practical case, inhomogeneity grows to cause the formation of the shear band where free volume concentration is higher initially. Shear strain inside shear band grows rapidly and can reach a critical value of strain which may cause crack initiation leading to full fracture.

Acknowledgements

The author greatly acknowledges the guidance of Prof. Tanmay K. Bhandakkar from Indian Institute of Technology of Bombay, Mumbai, India. This work would not be possible without the helpful discussions with him.

Abbreviations

MGs	metallic glasses
BMGs	bulk metallic glasses
STZ	shear transformation zone
1D	one-dimensional
2D	two-dimensional
ODE	ordinary differential equation
PDE	partial differential equation

Author details

Shank S. Kulkarni
The University of North Carolina at Charlotte, Charlotte, USA

*Address all correspondence to: skulka17@uncc.edu;
shankkulkarni1316@gmail.com

IntechOpen

References

[1] Turnbull D. Formation of crystal nuclei in liquid metals. Journal of Applied Physics. 1950;**21**(10):1022-1028

[2] Turnbull D, Cech RE. Microscopic observation of the solidification of small metal droplets. Journal of Applied Physics. 1950;**21**(8):804-810

[3] Jun WK, Willens RH, Duwez PO. Non-crystalline structure in solidified gold-silicon alloys. Nature. 1960; **187**(4740):869

[4] Telford M. The case for bulk metallic glass. Materials Today. 2004;**7**(3):36-43

[5] Ashby MF, Greer AL. Metallic glasses as structural materials. Scripta Materialia. 2006;**54**(3):321-326

[6] Xu J, Ramamurty U, Ma E. The fracture toughness of bulk metallic glasses. JOM. 2010;**62**(4):10-18

[7] Lewandowski JJ, Wang WH, Greer AL. Intrinsic plasticity or brittleness of metallic glasses. Philosophical Magazine Letters. 2005; **85**(2):77-87

[8] Murali P, Guo TF, Zhang YW, Narasimhan R, Li Y, Gao HJ. Atomic scale fluctuations govern brittle fracture and cavitation behavior in metallic glasses. Physical Review Letters. 2011; **107**(21):215501

[9] Gilman JJ. Flow via dislocations in ideal glasses. Journal of Applied Physics. 1973;**44**(2):675-679

[10] Gilman JJ. Mechanical behavior of metallic glasses. Journal of Applied Physics. 1975;**46**(4):1625-1633

[11] Chaudhari P, Levi A, Steinhardt P. Edge and screw dislocations in an amorphous solid. Physical Review Letters. 1979;**43**(20):1517

[12] Shi LT. Dislocation-like defects in an amorphous Lennard-Jones solid. Materials Science and Engineering. 1986;**81**:509-514

[13] Takeuchi S, Edagawa K. Atomistic simulation and modeling of localized shear deformation in metallic glasses. Progress in Materials Science. 2011; **56**(6):785-816

[14] Argon AS. Plastic deformation in metallic glasses. Acta Metallurgica. 1979; **27**(1):47-58

[15] Cohen MH, Turnbull D. Molecular transport in liquids and glasses. The Journal of Chemical Physics. 1959;**31**(5): 1164-1169

[16] Polk DE, Turnbull D. Flow of melt and glass forms of metallic alloys. Acta Metallurgica. 1972;**20**(4):493-498

[17] Spaepen F. A microscopic mechanism for steady state inhomogeneous flow in metallic glasses. Acta Metallurgica. 1977;**25**(4):407-415

[18] Taub AI, Spaepen F. The kinematics of structural relaxation of metallic glass. Acta Metallurgica. 1980;**28**(12):1781-1788

[19] Steif P, Spaepen F, Hutchinson J. Strain localization in amorphous metals. Acta Metallurgica. 1982;**30**(2):447-449

[20] Vaks VG. Possible mechanism for formation of localized shear bands in amorphous alloys. Physics Letters A. 1991;**159**(3):174-178

[21] Huang R, Suo Z, Prevost JH, Nix WD. Inhomogeneous deformation in metallic glasses. Journal of the Mechanics and Physics of Solids. 2002; **50**(5):1011-1027

[22] Kulkarni SS, Bhandakkar TK. Study of the effect of large deformation

through a finite deformation based constitutive model for metallic glasses. In: ASME 2018 International Mechanical Engineering Congress and Exposition; American Society of Mechanical Engineers; 2018. pp. V009T12A010-V009T12A010

[23] Gao YF. An implicit finite element method for simulating inhomogeneous deformation and shear bands of amorphous alloys based on the free-volume model. Modelling and Simulation in Materials Science and Engineering. 2006;**14**(8):1329

[24] De Hey P, Sietsma J, Van Den Beukel A. Structural disordering in amorphous Pd40Ni40P20 induced by high temperature deformation. Acta Materialia. 1998;**46**(16):5873-5882

[25] Yang B, Liaw PK, Wang G, Morrison M, Liu CT, Buchanan RA, et al. In-situ thermographic observation of mechanical damage in bulk-metallic glasses during fatigue and tensile experiments. Intermetallics. 2004;**12** (10–11):1265-1274

[26] Yang B, Liu CT, Nieh TG, Morrison ML, Liaw PK, Buchanan RA. Localized heating and fracture criterion for bulk metallic glasses. Journal of Materials Research. 2006;**21**(4):915-922

[27] Lewandowski JJ, Greer AL. Temperature rise at shear bands in metallic glasses. Nature Materials. 2006; 5(1):15

[28] Gao YF, Yang B, Nieh TG. Thermomechanical instability analysis of inhomogeneous deformation in amorphous alloys. Acta Materialia. 2007;**55**(7):2319-2327

[29] Jiang MQ, Dai LH. On the origin of shear banding instability in metallic glasses. Journal of the Mechanics and Physics of Solids. 2009;**57**(8):1267-1292

[30] Ruan HH, Zhang LC, Lu J. A new constitutive model for shear banding instability in metallic glass. International Journal of Solids and Structures. 2011;**48**(21):3112-3127

[31] Conner RD, Johnson WL, Paton NE, Nix WD. Shear bands and cracking of metallic glass plates in bending. Journal of Applied Physics. 2003;**94**(2):904-911

[32] Ravichandran G, Molinari A. Analysis of shear banding in metallic glasses under bending. Acta Materialia. 2005;**53**(15):4087-4095

[33] Dasgupta R, Hentschel HG, Procaccia I. Microscopic mechanism of shear bands in amorphous solids. Physical Review Letters. 2012;**109**(25): 255502

[34] Singh I, Guo TF, Murali P, Narasimhan R, Zhang YW, Gao HJ. Cavitation in materials with distributed weak zones: Implications on the origin of brittle fracture in metallic glasses. Journal of the Mechanics and Physics of Solids. 2013;**61**(4):1047-1064

[35] Huang X, Ling Z, Dai LH. Cavitation instabilities in bulk metallic glasses. International Journal of Solids and Structures. 2013;**50**(9):1364-1372

[36] Schuh CA, Hufnagel TC, Ramamurty U. Mechanical behavior of amorphous alloys. Acta Materialia. 2007;**55**(12):4067-4109

[37] Greer AL, Cheng YQ, Ma E. Shear bands in metallic glasses. Materials Science & Engineering R: Reports. 2013; **74**(4):71-132

[38] Timoshenko SP. Strength of Materials. New York: D. Van Nostrand Company; 1955

[39] Khan AS, Huang S. Continuum Theory of Plasticity. New York: John Wiley and Sons; 1995

[40] Yavari AR, Pang S, Louzguine-Luzgin DV, Inoue A, Lupu N, Nikolov N, et al. Change in thermal expansion coefficient of bulk metallic glasses Tg measured by real-time diffraction using high-energy synchrotron light. Journal of Metastable and Nanocrystalline Materials. 2003;**15**: 105-110

[41] Kato H, Chen HS, Inoue A. Relationship between thermal expansion coefficient and glass transition temperature in metallic glasses. Scripta Materialia. 2008;**58**(12): 1106-1109

[42] Spaepen F. Metallic glasses: Must shear bands be hot? Nature Materials. 2006;**5**(1):7

[43] Kulkarni SS, Bhandakkar TK. Thermo-mechanical model for inhomogeneous deformations in metallic glasses. Computer Aided Engineering. 2013:446-451

[44] Kulkarni SS. Modelling of shear bands in metallic glasses. Indian Institute of Technology Bombay. 2013

[45] Reddy JN. An Introduction to the Finite Element Method. New York: McGraw-Hill; 1993

[46] Hutton D. Fundamentals of Finite Element Analysis. New York: McGraw-Hill; 2004